Chromatographic Characterization of Polymers

ADVANCES IN CHEMISTRY SERIES 247

Chromatographic Characterization of Polymers

Hyphenated and Multidimensional Techniques

Theodore Provder, EDITOR
The Glidden Company
(Member of ICI Paints)

Howard G. Barth, EDITOR
DuPont

Marek W. Urban, EDITOR
North Dakota State University

American Chemical Society, Washington, DC 1995

Library of Congress Cataloging-in-Publication Data

Chromatographic characterization of polymers: hyphenated and multidimensional techniques / Theodore Provder, editor, Marek W. Urban, Editor, Howard G. Barth, editor.

p.   cm.—(Advances in chemistry series, ISSN 0065-2393; 247).

Includes bibliographical references (p. —) and index.

ISBN 0-8412-3132-X (case)

1. Polymers—Analysis.   2. Chromatographic analysis.

I. Provder, Theodore, 1939-   . II. Urban, Marek W., 1953-   . III. Barth, Howard G.   IV. Series.

QD1.A355   no. 247
[QD139.P6]
540 s—dc20   95-22137
[668.9]   CIP

The paper used in this publication meets the minimum requirements of American National Standard for Information Sciences—Permanence of Paper for Printed Library Materials, ANSI Z39.48-1984.   ∞

PRINTED IN THE UNITED STATES OF AMERICA

QD
I
.A355
NO 247

1995 Advisory Board

Advances in Chemistry Series

FOREWORD

The ADVANCES IN CHEMISTRY SERIES was founded in 1949 by the American Chemical Society as an outlet for symposia and collections of data in special areas of topical interest that could not be accommodated in the Society's journals. It provides a medium for symposia that would otherwise be fragmented because their papers would be distributed among several journals or not published at all.

Papers are reviewed critically according to ACS editorial standards and receive the careful attention and processing characteristic of ACS publications. Volumes in the ADVANCES IN CHEMISTRY SERIES maintain the integrity of the symposia on which they are based; however, verbatim reproductions of previously published papers are not accepted. Papers may include reports of research as well as reviews, because symposia may embrace both types of presentation.

ABOUT THE EDITORS

 THEODORE PROVDER is principal scientist at the Glidden Company's Research Center and is responsible for the research activities of the Materials Science and Analytical Services Department. He received a B.S. degree in chemistry from the University of Miami in 1961 and a Ph.D. inphysical chemistry from the University of Wisconsin in 1965.

After receiving his doctorate, he joined the Monsanto Company in St. Louis as a senior research chemist and carried out research on the characterization and material properties of exploratory polymers and composites. While he was at Monsanto, his research interests focused on molecular weight characterization, particularly by size-exclusion chromatography. Recently, his research has focused on size-exclusion chromatography, particle size distribution analysis, cure chemistry and physics, and the application of computers in the polymer laboratory. He is the author of more than 100 publications, is credited with three patents, and has edited or co-edited 10 volumes in the ACS Symposium Series and co-edited two volumes in the Advances in Chemistry series.

He was chairman of the ACS Division of Polymeric Materials: Science and Engineering, Inc., and has served on the advisory board of the ACS Books Department and the editorial advisory board for ACS's *Industrial & Engineering Chemistry Product Research and Development* journal. He is a member of the editorial boards for the *Journal of Coatings Technology* and *Progress in Organic Coatings*. He is also the treasurer for the Joint Polymer Education Committee of the Divisions of Polymeric Materials: Science and Engineering, Inc. and Polymer Chemistry, Inc.

He is a recipient of an SCM Corporation Scientific and Technical Award in the area of computer modeling and two Glidden Awards for Technical Excellence for advanced latex particle size analysis methods and instrumentation development. In addition, he received the coatings industry's highest honor by being awarded the 1987 Joseph J. Mattiello Lecture at the annual meeting of the Federation of Societies for Coatings Technology. In 1989, he was awarded the ACS Division of Polymeric

Materials: Science and Engineering, Inc., Roy W. Tess Award in Coatings. In 1993, Provder received the University of Missouri—Rolla Coatings Institute Distinguished Scientist Award in recognition of distinguished work in the field of coatings and polymer science.

HOWARD G. BARTH is a senior research associate of the Corporate Center for Analytical Sciences at the DuPont Experimental Station in Wilmington, Delaware. Before joining the DuPont Company in 1988, he was a research scientist and group leader at Hercules Research Center. He received his B.A. in 1969 and his Ph.D. in 1973 in analytical chemistry from Northeastern University. His specialties include polymer characterization, size-exclusion chromaography, and high-performance liquid chromatography. He has published more than 70 papers in these and related areas. Barth also edited the book *Modern Methods of Particle Size Analysis* (Wiley, 1984) and co-edited *Modern Methods of Polymer Characterization* (Wiley, 1991). He has edited five volumes on polymer characterization published in the *Journal of Applied Polymer Science* and co-edited two volumes in the ACS Symposium Series. Barth was on the instrumentation advisory panel of *Analytical Chemistry* and was associate editor of the *Journal of Applied Polymer Science.* He is cofounder and chairman of the International Symposium on Polymer Analysis and Characterization. He was recently appointed editor-in-chief of the *International Journal of Polymer Analysis and Characterization* (Gordon and Breach). Barth is past chairman of the Delaware Section of the American Chemical Society, and he presently serves as councillor. Barth is a member of the ACS Divisions of Analytical Chemistry, Polymer Chemistry, Inc., and Polymeric Materials: Science and Engineering, Inc.; the Society of Plastics Engineers; and the Delaware Valley Chromatography Forum. He is also a fellow of the American Institute of Chemists and a member of Sigma Xi.

MAREK W. URBAN received his M.S. degree in chemistry from Marquette University in 1979, followed by his Ph.D. from Michigan Technological University in 1984. Before joining North Dakota State University in 1986, he spent two years as a research associate in the Department of Macromolecular Science at Case Western Reserve University in Cleveland.

He is the author of more than 180 research papers, numerous review articles, and several book chapters and books, mostly published in the ACS Advances in Chemistry series. He is the author of *Vibrational Spectroscopy of Molecules and Macromolecules of Surfaces* (Wiley-Interscience), and he edited two books in the Advances in Chemistry series and two books in the ACS Symposium Series. He is also the editor of the ACS book series "Polymer Surfaces and Interfaces", which is part of the Professional Reference Books. His research interests range from characterization of polymer networks and surfaces using FTIR spectroscopy to structure–property relationships at surfaces and interfaces, and from spectroscopic measurement of diffusion in polymer networks to nonequilibrium thermodynamics. His academic research group is involved in the studies of polymer surfaces and interfaces using vibrational spectroscopic methods. For his pioneering work on rheophotoacoustic spectroscopy, he was awarded the 1990 Megger's Award presented for the most outstanding paper published in the *Applied Spectroscopy Journal*. The award is given by the Federation for Analytical Chemistry and Spectroscopy Societies. He also is credited with two patents in this area. For five consecutive years, from 1986 to 1991, he was honored by the 3M Company with the Young Faculty Award. From 1987 to 1988, he served as chair for the Society for Applied Spectroscopy, Minnesota Chapter. He is an invited speaker at many international conferences, Gordon Research conferences, and industrial laboratories. He serves as a consultant to several chemical companies.

His involvement in ACS symposia began in 1991, when he co-chaired the International Symposium on Spectroscopy of Polymers in Atlanta, Georgia, from which *Structure–Property Relations in Polymers* (Advances in Chemistry Number 236) was derived. He co-chaired the 1993 Symposium on Hyphenated Techniques in Polymer Characterization, which was held in Chicago, Illinois, during the ACS National Meeting, and served as chair of the International Symposium on Polymer Spectroscopy, which was held in Washington, D.C., in 1994. He is also a lecturer in spectroscopy workshops offered by ACS. At North Dakota State University, he is chair of the Polymers and Coatings Department, director of the summer coatings science short courses, and director of the National Science Foundation's Industry–University Coatings Research Center at North Dakota State University.

CONTENTS

PREFACE

THE GLOBAL, OPERATIVE BUSINESS AND SOCIETAL DRIVING FORCES of the mid-1990s are causing polymer-related industries to focus strongly on their core businesses and technological competencies. This focusing has produced a more directed approach to product development and a significant change in corporate R&D culture. The product development process is no longer a sequential process from R&D to product introduction into the marketplace. Instead, the process is highly nonlinear, nonsequential, and iterative in order to speed up product innovation, product development, and market introduction. Improving the effectiveness of the R&D process must be done in conjunction with strong regard for safety, health, and environmental values; waste reduction; energy conservation; product quality; improved product–process–customer economics; the need to satisfy and delight the customer; and the need to improve shareholder value.

The polymer science and technology required to meet product and market needs in the context of improving the effectiveness of the R&D process are generating highly complex polymeric systems that may be blends or composites of a variety of materials. As a result, measurement of average properties is no longer adequate to characterize and elucidate the nature of such complex polymeric materials. A combination of polymer analytical and characterization techniques or multidimensional analytical approaches is required to provide a synergism of analytical and characterization information to establish structure, property, and morphology processing relationships that can form a knowledge bridge between polymerization mechanisms and end-use performance. Advances in instrumentation technology and the need for analytical and characterization information synergism have led to an increase in the development of hyphenated characterization techniques and their application to polymeric materials characterization.

This book covers some of the significant advances in hyphenated chromatographic separation methods for polymer characterization. Chromatographic separation techniques in this volume include size-exclusion chromatography, liquid chromatography, and field flow fractionation methods that are used in conjunction with information-rich detectors such as molecular size-sensitive or compositional-sensitive detectors or coupled in cross-fractionation modes.

The first section of this book focuses on general considerations concerning hyphenated polymer chromatographic separation methods. The second section focuses on the use of light scattering and viscometry molecular size sensitive detectors, the issue of multidetection calibration, and some unique applications of these detectors. The third section focuses on the analysis and elucidation of compositional heterogeneity in copolymers and blends by using cross-fractionation approaches with compositional and molecular size sensitive detectors.

We hope this book will encourage and catalyze additional activity and method development in hyphenated chromatographic separation methods for polymer characterization.

Acknowledgments

We are grateful to the authors for their effective oral and written communications and to the reviewers for their critiques and constructive comments. We gratefully acknowledge the ACS Divisions of Polymeric Materials: Science and Engineering, Inc., and Analytical Chemistry, and the Petroleum Research Fund of the ACS for their financial support of the symposium on which this book is based.

THEODORE PROVDER
The Glidden Company
ICI Paints in North America
16651 Sprague Road
Strongsville, OH 44136

HOWARD G. BARTH
DuPont
Central Research and Development
Corporate Center for Analytical Sciences
Experimental Station
P.O. Box 80228
Wilmington, DE 19880–0228

MAREK W. URBAN
Department of Polymers and Coatings
North Dakota State University
Fargo, ND 58105

June 8, 1995

General Considerations

Hyphenated Polymer Separation Techniques

Present and Future Role

Howard G. Barth

DuPont, Experimental Station, Wilmington, DE 19880-0228

An overview is presented on recent developments in the use of hyphenated multidimensional separation and detection techniques for the characterization of polymeric materials. Emphasis has been placed on the use of on-line molecular-weight-sensitive detectors for size-exclusion chromatography (SEC). These detection systems are based on measuring Rayleigh light-scattering or intrinsic viscosity of the eluting polymer. With these types of detectors, one can determine absolute molecular weights as well as branching, molecular size, and polymer conformation as a function of molecular weight, without the use of column calibration. The determination of compositional heterogeneity using SEC with on-line selective detectors, such as UV, Fourier transform infrared, mass spectrometry, NMR, and even Raman spectrometry, is now being investigated. Multidimensional hyphenated techniques, such as orthogonal chromatography, temperature-rising elution fractionation–SEC, and SEC–high-performance liquid chromatography, are briefly discussed.

P OLYMERS ARE TYPICALLY COMPLEX MIXTURES in which the composition depends on polymerization kinetics and mechanism and process conditions. As we enter the twenty-first century, polymeric materials are becoming even more complex, consisting of polymer blends, composites, and branched and grafted structures of unusual architecture. To obtain polymeric materials of desired characteristics, polymer processing must be carefully controlled and monitored. Furthermore, we need to understand the influence of molecular parameters on polymer properties and end-use performance. As a result, we are faced with unprecedented

0065–2393/95/0247–0003$12.00/0

analytical challenges: molecular weight distributions (MWDs) and average chemical composition may no longer provide sufficient information for process and quality control nor define structure–property relationships. Methodologies to measure distributive properties based on chemical composition, branching, comonomer sequence distribution, and tacticity may also be required. More sophisticated analytical approaches are needed also for the characterization of macromonomers and telechelic oligomers used for polymer synthesis, especially in the coating industry.

Modern characterization methods for polymeric systems now require multidimensional analytical approaches rather than average properties of the whole sample. To meet these challenges, hyphenated methodologies are now emerging in which polymer separation techniques are being coupled to information-rich detectors or are being interfaced to a second chromatographic system, an approach referred to as cross-fractionation or two-dimensional (2D) separation. When seemingly incongruous techniques are interfaced in such a manner, one gets a dramatic increase in information content and a significant reduction of analysis time. This chapter presents an overview of the present status and future direction of hyphenated polymer separation techniques.

Size-Exclusion Chromatography

Size-exclusion chromatography (SEC) is the premier polymer characterization method for determining MWDs. As discussed in this volume and summarized in the following section, by hyphenating SEC with selective detectors, one can, in principle, completely characterize a polymer in terms of its molecular parameters and chemical composition in the time it takes to do a typical SEC analysis.

Molecular-Weight-Sensitive Detectors. In SEC, the separation mechanism is based on molecular hydrodynamic volume. For homopolymers, condensation polymers, and strictly alternating copolymers, there is a correspondence between elution volume and molecular weight; thus, chemically similar polymer standards of known molecular weight can be used for calibration. However, for SEC of random and block copolymers and branched polymers, no simple correspondence exists between elution volume and molecular weight because of possible compositional heterogeneity of these materials; as a result, molecular weight calibration with polymer standards can introduce a considerable amount of error. To address this problem, molecular-weight-sensitive detectors, based on Rayleigh light-scattering and intrinsic viscosity measurements, have been introduced (1).

In SEC-light scattering, a low-angle or multiangle light-scattering photometer is interfaced to the output of an SEC column. In most SEC experiments, the second virial coefficient can be neglected because of low polymer concentration, and the weight-average molecular weight, M_w, at each elution volume increment, i, can be determined from

$$M_{w,i} = R_{\theta,i}/Kc_iP(\theta)_i \qquad (1)$$

where K is the optical constant, c is the polymer concentration, R_θ is the Rayleigh ratio, and $P(\theta)$ is the particle scattering function. If a low-angle light-scattering instrument is used, $P(\theta)$ is close to unity and the M_w at each elution volume increment can be calculated directly. By assuming that each measured elution volume increment is monodisperse, that is, $M_{w,i} = M_{n,i}$, then all the statistical M_w averages can be calculated.

If a multiangle light-scattering instrument is used, the mean-square radius of gyration $\langle R_g^2 \rangle_i$ at each elution volume can also be obtained from the particle scattering function

$$1/P(\theta)_i = 1 + (4\pi/\lambda)^2 \sin^2 (\theta/2)\langle R_g^2 \rangle_i/3 \qquad (2)$$

In practice, however, the radius of gyration can only be determined for molecules >20 nm in diameter; below this size, angular dissymmetry is too low to measure precisely. By measuring radius of gyration as a function of M_w, insight into the molecular conformation of the polymer can be obtained.

With SEC-light scattering, absolute MWDs can be determined without column calibration. Furthermore, the branching distribution as a function of M_w can be determined using equation 3, if a multiangle light-scattering detector is used and the radius of gyration is >10 nm:

$$g_i = \langle R_g^2 \rangle_{b,i}/\langle R_g^2 \rangle_{l,i} \qquad (3)$$

where g_i is the radius of gyration branching factor and subscripts l and b signify the corresponding linear and branched polymers of the same M_w (2). If a low-angle light-scattering detector is used, the intrinsic viscosity branching factor g_i' can be used:

$$g_i' = (M_l/M_b)_i^{a+1} \qquad (4)$$

where subscripts l and b signify the corresponding linear and branched polymers eluting at the same elution volume and a is the Mark–Houwink exponent of the linear polymer (3).

For on-line intrinsic viscosity detection, the pressure drop across a capillary attached to the outlet of the SEC column is monitored. The ratio of the pressure drop of an eluting polymer ΔP to that of the mobile phase alone ΔP_o is equal to the relative viscosity η_{rel} of the sample:

$$\eta_{rel} = \Delta P/\Delta P_o \qquad (5)$$

Because intrinsic viscosity $[\eta]$ is related to relative viscosity

$$[\eta] = (\eta_{rel} - 1)/c \qquad (6)$$

where c is polymer concentration extrapolated to infinite dilution, then $[\eta]$ can be determined at each elution volume increment i:

$$[\eta]_i = (\Delta P_i/\Delta P_o) - 1)/c_i \qquad (7)$$

The value of c_i is measured using a concentration-sensitive detector, which is usually a differential refractometer. As in the case of light scattering, polymer concentration during an SEC experiment is considered to be close to infinite dilution; thus, extrapolation to zero concentration is not required. Providing that universal calibration is valid for a given SEC system, M_w data of the polymer at each elution volume increment M_i can be determined:

$$M_i = HV_i/[\eta]_i \qquad (8)$$

where HV is the corresponding hydrodynamic volume obtained from a universal calibration curve.

With the use of an on-line viscometer and universal calibration, "absolute" MWDs can be obtained. Because both M_w and intrinsic viscosity are known, the Mark–Houwink coefficients a and K can be calculated for the injected polymer:

$$[\eta] = KM^a \qquad (9)$$

The value of a can be used to determine macromolecular chain conformation and the presence of branching. The intrinsic viscosity branching factor can also be used to study branching as a function of M_w:

$$g_i' = ([\eta]_l/[\eta]_b)_{M,i} \qquad (10)$$

Furthermore, $R_{g,i}$ of a linear polymer at each elution volume increment can be calculated from the Flory–Fox equation (4):

$$R_{g,i} = 1/\sqrt{6}(M_i[\eta]_i/\Phi)^{1/3} \qquad (11)$$

where Φ is the Flory viscosity function.

Recently, both on-line light-scattering and viscosity detectors are being used together to give absolute M_w and intrinsic viscosity distributions without the need of universal calibration (1). When both detectors are used, one can determine precisely and accurately the hydrodynamic radius distribution (eq 8), the intrinsic viscosity branching factor distribution (eq 10), Mark–Houwink coefficients (eq 9), and the radius of gyration distribution of linear polymers (eq 11) in a single SEC experiment.

On-Line Spectroscopic Detectors. Depending on comonomer reactivity ratios and polymerization conditions, the chemical composition of copolymers can vary as a function of M_w; this distributive property is sometimes referred to as chemical drift or compositional heterogeneity of the first kind (Figure 1) (5). By hyphenating SEC with a selective detector, together with a concentration-sensitive detector, compositional heterogeneity can be tracked. For example, a UV detector set at a single wavelength can be used to monitor the chemical drift of a copolymer containing a chromophoric comonomer. If one is trying to characterize the nature of unknown polymer end groups or perhaps unknown polymer blends, UV-photodiode array detection would be the detector of choice.

There has been considerable interest in applying mass spectrometry (MS) (6) and Fourier-transform infrared spectrophotometry (FT-IR) (7) as on-line SEC detectors; these hyphenated systems show great potential for determining compositional heterogeneity of copolymers, blends, and oligomers. To eliminate interfering IR spectra caused by the mobile phase, a commercial evaporative interface device has been developed (8, 9) in which solvent is evaporated by nebulization of the column effluent onto a rotating aluminum-backed germanium disk. The disk is then transferred to an optical reader attached to an FT-IR spectrophotometer. Other selective on-line SEC detectors include the use of a conductivity detector to determine the charge distribution of polysaccharides as a function of M_w (10, 11), and an FT-Raman spectrometer for measuring microstructural variations in polybutadiene (12). It is of interest to note that although on-line UV, FT-IR, and MS detection systems are emerging technologies, there are surprisingly few reported studies describing on-line NMR systems (13), ostensibly because of the

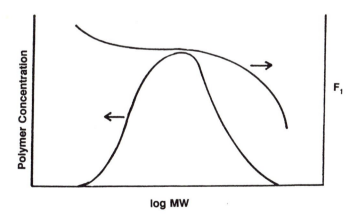

Figure 1. Chemical heterogeneity of the first kind in which F_1, the fraction of monomer 1 in a copolymer, varies as a function of M_w. (Reproduced from reference 5. Copyright 1986 American Chemical Society.)

need for much larger sample amounts. The information content that could be obtained from such a system, however, makes NMR detection a deserving area of research.

Polymer Cross-Fractionation

In addition to chemical heterogeneity of the first kind, another type of compositional heterogeneity may exist: chemical heterogeneity of the second kind, in which polymers of different composition but similar hydrodynamic volumes coelute (Figure 2) (5). In this situation, components have to be resolved using a second separation method, a process called cross-fractionation (14) or 2D separation, commonly referred to as column switching when done on-line. Although not commonly practiced, it is possible to use an automated injection system to collect and then divert a given SEC fraction into a high-performance liquid chromatography (HPLC) column for separation based on chemical composition. However, because of the complexity of these systems, fractions usually are collected and analyzed chromatographically off-line. Furthermore, chromatographic cross-fractionation can also be carried out by collecting fractions first from an HPLC column and then injecting them into an SEC column in the second dimension.

Orthogonal Chromatography. An interesting variation of online chromatographic fractionation is a technique called orthogonal chromatography (15). This method is essentially a hyphenated SEC–

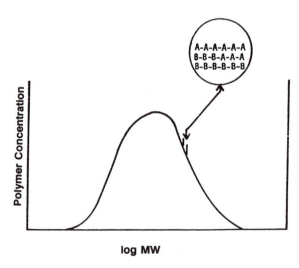

Figure 2.   Chemical heterogeneity of the second kind in which polymers of different composition may coelute because of similar hydrodynamic volumes. (Reproduced from reference 5. Copyright 1986 American Chemical Society.)

SEC technique, in which polymers are separated in an SEC column using a good solvent and fractions are then switched into a second SEC column using a poorer solvent as the mobile phase. If compositional heterogeneity is present, components that have similar molecular hydrodynamic volumes in the good solvent may have different hydrodynamic volumes in the poorer solvent and thus will be size separated in the second column. Additionally, adsorption onto the packing may occur in the second dimension, giving rise to selective retention of components.

Temperature-Rising Elution Fractionation.   In temperature-rising elution fractionation (TREF), a polymer is dissolved at an elevated temperature in a good solvent and then injected into a chromatographic column packed with an inert support (*16*). The flow rate is then turned off and the column slowly cooled, typically 2–10 °C/h. The higher melting fractions crystallize first, followed by less crystalline material, thus forming a "layered" structure on the inert support. The flow rate is then turned on and the temperature is increased, thus reversing the process: the less crystalline components elute first followed by more crystalline material. Thus, separation is based on polymer crystallinity imposed by chemical compositional heterogeneity, as well as by architectural heterogeneities, that is, short-chain branching, tacticity, or comonomer sequence distribution. Because this method is independent of M_w, providing that the M_w is >10,000 g/mol (*17*), superposition of M_w on the chemical composition distribution is not a major concern.

To cross-fractionate a given TREF fraction in terms of MWD, on-line TREF–SEC–FT-IR instrumentation has been developed to obtain three-dimensional plots of polymer concentration, M_w, and composition (short-chain branching) (*18*).

Future Challenges and Directions

Hyphenated multidimensional analytical instrumentation requires careful calibration and maintenance to obtain high quality, meaningful data (*19*). Because of the propagation of systematic and random errors as different analytical instrumentation are interfaced, frequent calibration using well-characterized polymer standards is required even for absolute M_w-sensitive detectors. Furthermore, the relatively low signal-to-noise ratio at the ends of the MWD can lead to significant uncertainties in these regions of the distribution; unfortunately, these areas of the distribution can profoundly affect polymer properties.

Fundamental limitations in SEC, most notably imperfect resolution and lack of truly universal concentration-sensitive detectors, also may add to the uncertainty of hyphenated instrumental methods. Furthermore, there are still difficult separation problems to be solved, for example, branching heterogeneity of the second kind (Figure 3), especially for long-chain branched structures, in which polymers of different architecture coelute because of similar hydrodynamic volumes. Perhaps on-line detection systems based on viscoelastic behavior may help out here.

Data interpretation and processing can be complicated. In addition, software to handle hyphenated methodologies are still under development, and many laboratories find it more convenient to write their own software. As a result, most hyphenated polymer separation instrumentation are limited presently to research laboratories rather than to plant environments. Continued developments in digital electronics, laser-based detector technology, and computer data acquisition and processing will result eventually in easy to use, automated hyphenated instrumentation for process and quality control.

Present emphasis is being placed on the use of detector combinations; in fact, the "triple-detection system", SEC–light-scattering-viscometry–differential refractometry, is being used more frequently, sometimes with the addition of an on-line UV detector to monitor composition (20).

On-line MS detectors for SEC offer great promise for serving simultaneously as an absolute M_w detector, concentration-sensitive detector (i.e., total ion current), and composition detector, especially for oligomers and low-MW polymers. Successful developments in the use of matrix-assisted laser desorption ionization (MALDI) MS for high-MW polymers are extremely encouraging, and on-line MALDI MS should

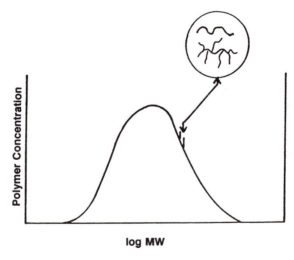

Figure 3. Chemical heterogeneity of the second kind in which polymers of different architecture may coelute because of similar hydrodynamic volumes.

soon follow. On-line NMR for SEC offers a great deal of promise, and it is hoped that we will see some growth in this area. By stringing together different selective detectors and connecting them to chromatographic systems based on chemical composition and MW separations, complete characterization of complex polymeric materials may be achievable in a single experiment. This information is critically needed to establish structure–property–processing relationships to tailor materials of given properties and end-use applications.

References

1. Jackson, C.; Barth, H. G. In *Molecular Weight Sensitive Detectors for Size Exclusion Chromatography*; Wu, C.-S., Ed.; Dekker: New York, 1995; Chapter 4, pp 103–145.
2. Zimm, B. H.; Stockmayer, W. H. *J. Chem. Phys.* **1949**, *17*, 1301.
3. Rudin, A. In *Modern Methods of Polymer Characterization*; Barth, H. G.; Mays, J. W., Eds.; Wiley: New York, 1991; p 103.
4. Flory, P. J.; Fox, T. G. *J. Am. Chem. Soc.* **1951**, *73*, 1904.
5. Barth, H. G. In *Water-Soluble Polymers: Beauty with Performance*; Glass, J. E., Ed.; Advances in Chemistry 213; American Chemical Society: Washington, DC, 1986; pp 31–55.
6. Prokai, L.; Simonsick, W. J., Jr. *Rapid Commun. Mass Spectrom.* **1993**, *7*, 853.
7. Nishikida, K.; Housaki, T.; Morimoto, M.; Kinoshita, T. *J. Chromatogr.* **1990**, *517*, 209.
8. Cheung, P.; Balke, S. T.; Schunk, T. C.; Mourey, T. H. *J. Appl. Polym. Sci., Appl. Polym. Symp. Ed.* **1993**, *52*, 105.
9. Wheeler, L. M.; Willis, J. N. *Appl. Spectrosc.* **1993**, *47*, 1128.
10. Rinaudo, M.; Danhelka, J.; Milas, M. *Carbohydr. Polym.* **1993**, *21*, 1.
11. Ploeger, A. *J. Food Sci.* **1992**, *57*, 1185.
12. Edwards, H. G. M.; Johnson, A. F.; Lewis I. R. *J. Raman Spectrosc.* **1993**, *24*, 435.
13. Ute, K.; Hatada, K. *Anal. Sci.* **1991**, *7*, 1629.
14. Glöckner, G.; Barth, H. G. *J. Chromatogr.* **1990**, *499*, 645.
15. Balke, S. T. In *Detection and Data Analysis in Size Exclusion Chromatography*; Provder, T., Ed.; ACS Symposium Series 352; American Chemical Society: Washington, DC, 1987; pp 59–77.
16. Wild, L. *Adv. Polym. Sci.* **1990**, *98*, 1.
17. Wild, L.; Ryle, D.; Knobeloch, D.; Peat, I. *J. Polym. Sci., Polym. Phys. Ed.* **1982**, *20*, 441.
18. Usami, T.; Gotoh, Y.; Umemoto, H.; Takayama, S. *J. Appl. Polym. Sci., Appl. Polym. Symp. Ed.* **1993**, *52*, 145.
19. Jackson, C.; Barth, H. G. Chapter 5 in this volume.
20. Barth, H. G.; Boyes, B. E.; Jackson, C. *Anal. Chem.* **1994**, *66*, 595R.

RECEIVED for review January 6, 1994. ACCEPTED revised manuscript December 12, 1994.

Limiting Conditions in the Liquid Chromatography of Polymers

David J. Hunkeler,[1] Miroslav Janco,[2] Valeria V. Guryanova,[3] and Dusan Berek[2]

[1] Department of Chemical Engineering, Vanderbilt University, Nashville, TN 37235
[2] Polymer Institute, Slovak Academy of Sciences, Dubravska cesta 9, 842 36 Bratislava, Slovakia
[3] Research Institute of Plastic, Perovsky proezd 35, 111 112 Moscow, Russia

The use of "limiting conditions of solubility" in the characterization of polymer mixtures is discussed. This use involves a binary eluent that is a weak nonsolvent or a poor solvent for one of the components of the polymer blend. The polymers are, however, injected in a thermodynamically good solvent. These limiting conditions are so named because the separation involves the analysis of one solute component that is on the limit of its solubility. The limiting condition is achieved through a balance of entropic (size exclusion) and enthalpic (precipitation) effects. The method is, therefore, similar to the critical condition approach, which also involves the combination and balancing of size exclusion and an interactive mechanism. However, the critical condition approach involves adsorption chromatography in place of precipitation. Our experiments indicated that limiting conditions are characterized by retention volumes independent of the hydrodynamic volume for macromolecules as high as 1,000,000 Da, whereas critical conditions are limited to lower molar masses. The limiting conditions discussed include the characterization blends of polymers of various polarity over bare silica gels. An evaporative light-scattering detector is used to avoid problems of injection solvent and solute peak overlapping. Although the flow is isocratic, a microgradient does exist through the column.

A POLYMER CHARACTERIZATION METHOD based on liquid chromatographic measurements using binary eluents as a mobile phase is dis-

0065–2393/95/0247–0013$12.00/0

cussed (1, 2). In this approach, which was termed "limiting conditions of solubility", a set of conditions are identified in which a homopolymer with a molecular weight range of 10^2–10^6 Da excludes at the same retention volume, independent of the polymer molecular weight. These limiting conditions are specific to the polymer–eluent–sorbent system used. The limiting conditions are accomplished by using a mobile phase that is a poor solvent, or even a nonsolvent, for the polymer probe (mixture of a thermodynamically good solvent and a nonsolvent at the temperatures used in the measurements). However, the polymer is injected in a thermodynamically good solvent. In this case, homopolymers with different molecular weights leave from the chromatographic column at the same retention volume. This volume is roughly equal to the volume of liquid in column. The following mechanism is believed to cause the limiting condition phenomena.

At low levels of nonsolvent, such as methanol in toluene or water in tetrahydrofuran (THF) for the polystyrene (PS) or poly(methyl metacrylate) (PMMA) systems, the calibration curves shift slightly to lower retention volumes due to the influence of adsorption, partition, and a reduced pore size (3). At higher quantities of nonsolvent, in the vicinity of the θ-composition (e.g., 79.6% of toluene for PS in toluene–methanol at 25 °C) the thermodynamic quality of the solvent is strongly reduced. Mixtures containing more methanol are nonsolvents for PSs. If such a mixture is used as a size-exclusion chromatography (SEC) eluent and the polymer is dissolved in a good solvent (toluene), macromolecules move together with the zone of their initial solvent. If macromolecules move faster because of exclusion processes, they encounter the nonsolvent and precipitate. They then redissolve as the injection zone (pure solvent) reaches the precipitated polymer. This "microgradient" process of precipitation–redissolution occurs many times throughout the column with the polymer eluting just in the front part of the solvent zone (Figure 1). As a consequence, the macromolecules move with a velocity similar to the velocity of the solvent zone. This fact may be used in the analysis of polymer mixtures: limiting conditions are chosen in such a way that one component of polymer mixture is eluted in the full column volume, whereas the second one is normally characterized in the SEC mode.

The primary differences between limiting conditions of solubility and Belenkii's "critical condition of adsorption" approach (4, 5) are the use of a thermodynamically poor (bad) eluent, or even an eluent that is a nonsolvent for the polymer, whereas the polymer is dissolved and injected in a thermodynamically good solvent. The limiting condition

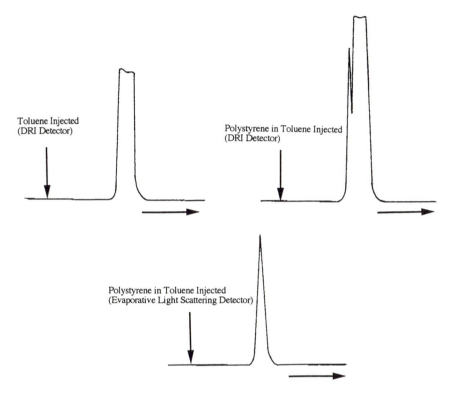

Figure 1. Chromatograms obtained with a 250 × 6-mm column packed with silica gel. The eluent was toluene–methanol at 68–32 vol%.

method offers an advantage relative to Belenkii's approach because it is experimentally less demanding:

- the sample dissolution is faster

- the determination of the appropriate eluent composition is easier

- limiting conditions are less sensitive to slight (minute) changes in the eluent composition and temperature as well as to the presence of impurities, particularly water traces in the eluent

- limiting conditions can be used for high molecular weight polymers (up to 1,000,000 g/mol), whereas critical conditions are, thus far, limited to the characterization of macromolecules up to 100,000 g/mol.

Both approaches are accomplished isocratically and therefore the problem of irreproducible gradient production, which is a limitation for the quantitative analysis of copolymers by gradient chromatography, is avoided.

Experimental Details

Chromatographic measurements were made on bare and modified silica gel sorbents prepared in the Laboratory of Liquid Chromatography of the Slovak Academy of Sciences (additional details are provided in reference 1). Bare silica gels, SGX-200, SGX-500, and SGX-1000, obtained from Tessek (Prague, Czech Republic) were also used. These gels were packed in 250-mm stainless steel columns with a 6 mm internal diameter A RIDK 102 differential refractive index detector and an HPP 4001 high pressure pump (both from Laboratory Instruments Company, Prague, Czech Republic) were used. A Waters 501 pump (Waters, Milford, MA) and a Cunow DDL-21 (Cunow, Cergy Pointoise, France) evaporative light-scattering detector were also employed. Pressure was measured with a custommade pressure gauge (0–25 MPa) (Institute of Chemical Process Fundamentals, Czechoslovak Academy of Sciences, Prague). The refractive index or scattered light signals and pressure signals were recorded on a type 185 two-pen chart recorder (Kutesz, Budapest, Hungary). The data were also collected on-line using a Waters PC Based Data Acquisition System. The injector was a PK-1 model (Institute of Chemical Process Fundamentals, Czechoslovak Academy of Sciences, Prague). Sample injections consisted of 10 mL of a polymer solution in a good solvent. The injected concentration was 1.0 mg/mL with the RIDK-102 detector and 0.5 mg/mL with DDL-21 detector system.

Narrow molecular weight distribution PS standards (polydispersity 1.06–1.20) were obtained from Pressure Chemicals Corporation. PMMA standards were obtained from Rohm and Haas (Darmstadt, Germany). Analytical grade solvents (toluene and methanol) were obtained from Lachema (Brno, Czech Republic) and used without further purification. THF was obtained from Merck (Darmstadt, Germany).

Results

In this chapter we discuss limiting conditions of solubility. These conditions were observed for polymers of varying polarity (PS and PMMA) using binary eluent mixtures that combined a polar nonsolvent with either a polar or nonpolar solvent.

The calibration curves for PS and PMMA in mixtures of toluene–methanol and THF–water are shown in Figures 2–5. These figures show that as the thermodynamic quality of the solvent is marginally reduced, the calibration curve shifts to lower retention volumes due to the combination of adsorption, partition, and a reduced pore size discussed earlier. At higher levels of nonsolvent, the calibration curve is very sensitive to the composition of the mobile phase. For compositions of toluene–methanol less rich in toluene than the θ composition, the calibration curve did not continue to shift to the left and began shifting to the right,

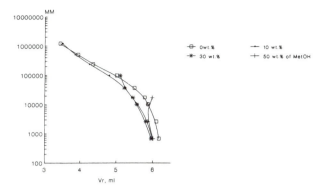

Figure 2.   Plot of the molar mass (MM; g/mol) as a function of the retention volume (mL). The calibration curves for narrow PS standards in a mixed eluent (toluene–methanol) at various compositions. The sorbent was bare silica gel with 100-nm pores.

that is, to higher retention volumes. This shift continues with increasing amounts of nonsolvent (methanol) in eluent. However, at a sufficiently high level of the nonsolvent content, or for high molar masses of polymer, the sample is retained within column packing and does not elute. For example, PS with a molar mass above 10^5 Da does not leave silica gel SGX-1000 if the eluent contains 30 wt% of methanol or more. This might, however, present no serious problem for a simultaneous SEC characterization of a second polymer, for example, PMMA. Indeed, our

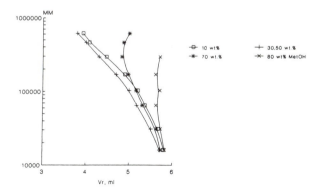

Figure 3.   Plot of the molar mass (MM; g/mol) as a function of the retention volume (mL). The calibration curves for narrow PMMA standards in a mixed eluent (toluene–methanol) at various compositions. The sorbent was bare silica gel with 100-nm pores.

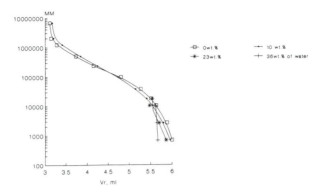

Figure 4.   Plot of the molar mass (MM; g/mol) as a function of the retention volume (mL). The calibration curves are for narrow PS standards in a mixed eluent (THF–water) at various compositions. The sorbent was bare silica gel with 100-nm pores.

measurements have revealed that the column packing is able to retain rather large amounts of adsorbed polymer without measurable changes in its SEC properties (6).

Table I summarizes the polymer–eluent–sorbent systems on which limiting conditions have been observed in our laboratories. It is evident that limiting conditions are observable on different systems of polymer–column packing–solvent:nonsolvent. These observation are leading us to believe that the limiting condition phenomenon is relatively generic.

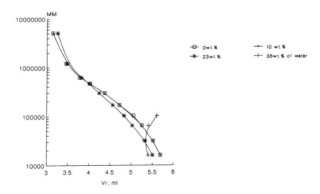

Figure 5.   Plot of the molar mass (MM; g/mol) as a function of the retention volume (mL). The calibration curves for narrow PMMA standards in a mixed eluent (THF–water) at various compositions. The sorbent was bare silica gel with 100-nm pores.

Table I.   Systems of Polymer–Sorbent–Eluent Where Limiting
Conditions Have Been Observed

Polymer	Sorbent	Detector	Eluent System	Eluent Components wt %/wt %
PS	unmodified silica gel, 10-μm particles, 800-Å pore size	differential refractive index	toluene–methanol	68/32
PS	unmodified silica gel, 10-μm particles, 1000-Å pore size	evaporative light scattering	toluene–methanol	50/50
PS	unmodified silica gel, 10-μm particles, 1000-Å pore size	differential refractive index, evaporative light scattering	THF–water	64/36
PMMA	unmodifed silica gel, 10-μm particles, 800-Å pore size	differential refractive index	toluene–methanol	27/73
PMMA	unmodified silica gel, 10-μm particles, 1000-Å pore size	evaporative light scattering	toluene–methanol	20/80
PMMA	unmodified silica gel, 10-μm particles; 1000-Å pore size	differential refractive index, evaporative light scattering	THF–water	64/36

Acknowledgments

We thank Ivan Novak (PI SAS) for preparing the silica gels used in this research. This research was supported by the International Division of the National Science Foundation (Washington, DC) grant INT–9208552.

References

1. Hunkeler, D.; Macko, T.; Berek, D. *Polym. Mater. Sci. Eng.* **1991,** 65, 101.
2. Hunkeler, D.; Macko, T.; Berek, D. In *Chromatography of Polymers: Characterization by SEC and FFF;* Provder, T., Ed.; ACS Symposium Series 521; American Chemical Society: Washington, DC, 1993; pp 90–102.
3. Berek, D.; Bakos, D.; Bleha, T.; Soltes, L. *Makromol. Chem.* **1975,** 176, 391.
4. Belenkii, B. G.; Gankina, E. S. *J. Chromatogr.* **1977,** 141, 13.
5. Skvortsov, A. M.; Gorbunov, A. A. *J. Chromatogr.* **1990,** 507, 487.
6. Janco, M.; Prudskova, T.; Berek, D. **1995,** 55, 393. *J. Appl. Polym. Sci.* **1995,** 55, 393.

RECEIVED for review January 14, 1994. ACCEPTED revised manuscript November 29, 1994.

Isoperichoric Focusing Field-Flow Fractionation Based on Coupling of Primary and Secondary Field Action

Josef Janča

Pôle Sciences et Technologie, Université de La Rochelle, 17042 La Rochelle
Cedex 01, France

*The spatially oriented gradient of the effective property of a carrier
liquid produced by a primary field coupled with the action of a
secondary field can generate the isoperichoric focusing of the dispersed species and separate them according to differences responding to the effective property gradient. This concept can be applied
under static conditions in thin-layer focusing or under dynamic
flow conditions in focusing field-flow fractionation. The gradient
is established by the effect of the primary field, and the isoperichoric
focused zones are formed by the coupled effect of the gradient and
of the primary or secondary field. The isoperichoric focusing theory
was developed to describe the particular processes operating in
focusing separations. Computer simulation was used to demonstrate
the potentials of the proposed principle, and few experiments were
performed under static and dynamic conditions.*

FIELD-FLOW FRACTIONATION (FFF) is a separation method convenient
for the analysis and characterization of macromolecules and particles of
synthetic or natural origin. Under the appropriate experimental conditions, it also can be applied for the preparative fractionation. The
separation is due to a simultaneous action of the effective field forces
and of the carrier liquid flow inside an open channel on the dissolved
or suspended macromolecules or particles. The carrier liquid flows in
the direction of the channel longitudinal axis and the field forces act in
the perpendicular direction across the channel thickness. Each component of the fractionated sample interacting with the field forces is
selectively transported across the channel. This concentrating process

induces an opposite diffusion flux. At the dynamic equilibrium, a quasi-stable concentration distribution of each sample component across the channel is established.

A suitable combination of the effective forces can give rise to the formation of the steady-state zones of the individual sample components focused across the channel at different lateral positions where the resulting force is zero (1). The shape of an individual focused zone in the direction of the focusing is described by a distribution function with the maximum concentration at the position where the focusing forces vanish. The focused zones are carried by the flow in the direction perpendicular to the focusing field action with linear velocities corresponding to their positions in the established flow velocity profile. As a result, the separation of different sample components occurs. This basic principle is shown schematically in Figure 1. Various combinations of the fields and gradients determining individual methods of focusing FFF were described in detail in references 2 and 3.

Field and Gradient Combinations

Effective Property Gradient of the Carrier Liquid Coupled with the Field Action (Isoperichoric Focusing). Focusing can appear as a consequence of the effective property gradient of the carrier liquid in the direction across the channel combined with the primary or secondary transversal field action. The density gradient in sedimentation-flotation focusing FFF or the pH gradient in isoelectric focusing FFF were already implemented (2, 3).

Preformed Gradient Combined with the Field Action. The effective property gradient of the carrier liquid can be preformed at the beginning of the channel and combined with the primary or secondary field forces. If, for example, the carrier liquids of various densities are pumped into the channel through several inlets under conditions hindering their mixing, the step density gradient is formed. The preforming is not limited to density gradient.

Cross-Flow Velocity Gradient Combined with the Field Action. The focusing effect can be achieved by the action of the gradient of linear flow velocity of the carrier liquid in the direction opposite to the action of transversal field forces. The longitudinal flow of the carrier liquid can be imposed simultaneously. This elutriation focusing FFF method was investigated experimentally by using the trapezoidal cross-section channel (*see* references 2 and 3 for review). In rectangular cross-section channel, the flow rates through the opposite semipermeable walls should be different, thus forming the flow velocity gradient across the channel.

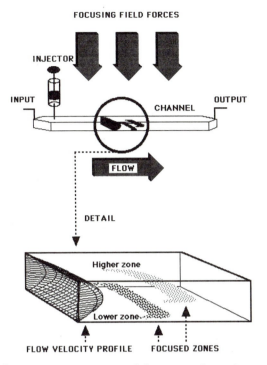

Figure 1.   Schematic representation of the isoperichoric focusing FFF concept.

Lift Forces Combined with the Field Action. The hydrodynamic lift forces that appear at high flow rates of the carrier liquid combined with the primary field are able to concentrate the hard suspended particles into the focused layers. The retention behavior of the particles under the simultaneous effect of the primary field and lift forces generated by the high longitudinal flow rate can vary with the nature of various applied primary field forces.

Shear Stress Combined with the Field Action. The high shear gradient at high flow rate of the carrier liquid leads to the deformation of the soft macromolecular chains. This deformation results in a decrease of the chain entropy. The established entropy gradient generates the driving forces that displace the macromolecules into a low-shear zone (4). A temperature gradient acting as a primary field generates the thermal diffusion flux of the macromolecules that opposes the flux due to the entropy changes generated motion and due to the diffusion. At a position where all the driving forces are balanced, the focusing of the sample components can appear (5).

Gradient of the Nonhomogeneous Field Action.   A nonhomogeneous high-gradient magnetic field formed in a cylindrical capillary with a concentric ferromagnetic wire placed in a homogeneous magnetic field can be used to separate various paramagnetic and diamagnetic particles of the biological origin by a mechanism of focusing FFF (6). A concentration of paramagnetic particles near the ferromagnetic wire and the focusing of diamagnetic particles in a free volume of the capillary should occur.

The focusing FFF can be used for continuous preparative fractionation (7, 8). If the fractionation channel is equipped with several outlet capillaries at various positions in the direction of focusing and the sample to be fractionated is continuously pumped into the channel, the focused layers eluting through the individual outlets can be collected.

The important advantage of the focusing FFF methods is that only a limited amount of the fractionated sample comes into direct contact with the walls of the separation channel. This is an important factor in fractionation of sensitive biological material.

Isoperichoric Focusing

The individual focusing FFF methods and techniques can exploit various mechanisms based on combination of the driving forces and gradients to establish the focused zones, and various experimental arrangements can be constructed to achieve the effective fractionation. This chapter reviews the theoretical and experimental achievements (although rather limited) concerning isoperichoric focusing under static conditions in thin-layer focusing (TLF) cells and under dynamic conditions in focusing FFF channels.

The spatially oriented gradient of the effective property of the carrier liquid combined with the field action induce the formation of the isoperichoric-focused zones of particulate dispersed species in general and under conditions of focusing FFF in particular (9). The term isoperichoric focusing was introduced by Kolin (10) and designates a condition in which a responding parameter of the focused sample component becomes equal to the corresponding effective property of the carrier liquid. This term overlays particular processes like isopycnic, isoelectric, and so on, focusing. Either the proper force of the primary field generating the gradient (1–3) or a secondary field of different nature coupled with the established gradient (11–14) can produce the focusing effect.

Theory of Focused Zone Formation.   The total flux of all species due to the effect of the fields and gradients can be written as a sum of the transversal flux of the carrier liquid modifier and of the focused sample component in the direction of x-coordinate. Although physically coupled, these fluxes can mathematically be treated as two independent processes. It holds at equilibrium for the flux of the carrier liquid modifier (m)

$$D_m \mathbf{r}_{m,x} \frac{\partial c_m}{\partial x} - \mathbf{U}_m c_m = 0 \qquad (1)$$

and for the flux of the focused sample component (i)

$$D_i \mathbf{r}_{i,x} \frac{\partial c_i}{\partial x} - \mathbf{U}_i(x) c_i = 0 \qquad (2)$$

where D is the diffusion coefficient, $\mathbf{U}$ and $\mathbf{r}$ are the velocity and unit vectors, and c is the concentration. The solution of equation 1 leads to the exponential concentration distribution of the modifier across the channel:

$$c_m(x) = \frac{c_{m,ave} w \, |U_m|}{D_m(1 - \exp(-w\,|U_m|/D_m))} \exp(-x\,|U_m|/D_m) \qquad (3)$$

where $c_{m,ave}$ is the average concentration of the modifier and w is the thickness of the channel or cell. The force $F_i(x)$, acting on a single particle or macromolecule undergoing the focusing, can be written in scalar form as

$$F_i(x) = - \left| \left(\frac{\mathrm{d} F_i(x)}{\mathrm{d} x} \right)_{x = x_{max,i}} \right| (x - x_{max,i}) \qquad (4)$$

By considering that the effective force gradient, $(\mathrm{d} F_i(x)/\mathrm{d} x)_{x=x_{max,i}}$, is constant near the focused zone position, $x_{max,i}$, the solution is (1)

$$c_i(x) = c_{i,max} \exp \left[-\frac{1}{2kT} \left| \left(\frac{\mathrm{d} F_i(x)}{\mathrm{d} x} \right)_{x=x_{max,i}} \right| (x - x_{max,i})^2 \right] \qquad (5)$$

which is the Gaussian distribution function for the sample component concentration profile in the direction of the focusing forces.

   The above concept was developed for general focusing FFF (1). The resulting Gaussian distribution function describing the shape of the concentration profile of the focused sample component is analogous to concentration distribution used to describe the zones in conventional centrifugal isopycnic focusing (15) or in isoelectric focusing (16). In both cases, the transversal gradient of the corresponding focusing forces, $(\mathrm{d} F_i(x)/\mathrm{d} x)_{x=x_{max,i}}$, is assumed to be constant within the range of the focused zone.

Giddings and Dahlgren (17) proposed the use of the Taylor's series to describe the driving focusing force around the focusing position. The transversal concentration distribution in FFF channel or the general shape of the focused zone in equilibrium gradient focusing is then described by a more complicated function (18).

This function formally does not impose the assumption of the constant gradient of the focusing forces; however, if truncated, it can be applied only in the near vicinity of the focusing position $x_{max,i}$, thus for narrow zones. By retaining only the first term of the series, this function is reduced to Gaussian distribution.

The situation becomes more complicated for broad focused zones that can appear, for example, because of lower intensity of the secondary focusing forces. In such a case, the Gaussian distribution function deviates substantially from the rigorous distribution function described in the following paragraph. Higher order approximations using Taylor's series are completely incoherent, exhibiting some oscillations and are only very slowly converging to the rigorous distribution function (9).

A completely different approach (11), which takes into account the actual shape of the established gradient of the focusing forces in the direction of x-coordinate, leads to purely analytic solution without a priori assumption of the constancy of the gradient in the domain of the focused zone (15, 16) and without approximation by a series (17, 18).

The transversal gradient of the focusing forces can be generated, for example, in a binary or multicomponent carrier liquid whose two or more components are affected unevenly by the primary homogeneous field. Various effective property gradients of the carrier liquid differing by their effect on the focused sample component can be exploited (2, 3). All of them occur because of the concentration distribution of the carrier liquid modifier. The concentration distribution of the modifier can be established due to the effect of the primary field forces, which can act also to form the focused zones of the separated sample components.

Another possibility is to superpose a secondary field of different nature on the gradient generated by the primary field to induce the focusing effect (11). The focusing force proper is induced by the coupled effect of the transversal gradient and of the primary or secondary field action.

The concept of spatially oriented gradient coupling of the effective property of the carrier liquid generated by the primary field with the action of the secondary field of different nature can be exemplified by the focusing of the particulate species in a density gradient by the effect of the natural gravitation, whereas the density gradient is formed by the action of the primary electric field on two-component colloidal carrier liquid. In such a case, the density gradient in the direction of the primary electric field action on charged density modifier colloid particles is described in references 12 and 13 by

$$\rho(x) = \rho_l + \frac{\phi_{m,ave}\Delta\rho_m w \,|\, U_{m,e}\,|\,/D_m}{(1 - \exp(-w\,|\,U_{m,e}\,|\,/D_m))}\exp(-x\,|\,U_{m,e}\,|\,/D_m) \qquad (6)$$

where $U_{m,e}$ is the velocity of the motion of the modifier particles due to the primary electric field, $\Delta\rho_m = \rho_m - \rho_l$, where ρ_m and ρ_l are the densities of the modifier and of the suspending liquid, respectively, and $\phi_{m,ave}$ is the average volume fraction of the modifier. The focusing force is given by

$$F_i(x) = (\rho(x) - \rho_i)v_i g_f \qquad (7)$$

where v_i is the volume of the focused particle and g_f is the acceleration of the focusing field (gravitation in this case). The final relationship is

$$c_i(x) = c_{i,max}\exp\left\{-\left[\frac{v_i g_f \phi_{m,ave}\Delta\rho_m w}{kT(1 - \exp(-w\,|\,U_{m,e}\,|\,/D_m))}\right] \times \left[\exp(-x\,|\,U_{m,e}\,|\,/D_m)\right.\right.$$
$$\left.\left. - \exp(-x_{max,i}\,|\,U_{m,e}\,|\,/D_m)\left(1 + \frac{|\,U_{m,e}\,|\,(x_{max,i} - x)}{D_m}\right)\right]\right\} \qquad (8)$$

Equation 8 is the rigorous concentration distribution function of an individual focused component of the fractionated sample.

A more general approach not related exclusively to the isopycnic focusing was published recently (9).

Resolution. An important quantitative parameter frequently used to evaluate the performance of the separation methods is the resolution, R_s. This parameter is calculated as the ratio of the relative distance of two totally or partially resolved zones to the average standard deviation of their widths expressed in the same units. The resolution of the focusing process that is described approximately by Gaussian distribution is given by

$$R_s = \frac{U_{m,e}}{2D_m}\sqrt{\frac{v_i g_f \phi_{m,ave}\Delta\rho_m w}{kT(1 - \exp(-w\,|\,U_{m,e}\,|\,/D_m))}}$$
$$\times \frac{|x_{max1} - x_{max2}|}{\dfrac{1}{\sqrt{\exp(-x_{max1}\,|\,U_{m,e}\,|\,/D_m)}} + \dfrac{1}{\sqrt{\exp(-x_{max2}\,|\,U_{m,e}\,|\,/D_m)}}} \qquad (9)$$

Computer Simulation. The model conditions were chosen to calculate the shapes of the focused zones by using the previously described rigorous distribution function, equation 8, and Gaussian distribution function, equation 5. The results are represented in Figure 2 as the shapes of several focused zones. The important conclusion from this model calculation is that the rigorous or Gaussian distribution functions are almost identical for narrow zones focused at lower $x_{max,i}$ values. However, the shapes of the focused zones are substantially different for broader zones focused at higher $x_{max,i}$ values if the calculation is performed either for rigorous or Gaussian distribution function. As concerns the resolution of the zones demonstrated in Figure 2, it does not change monotonously with increasing distance of the zones relative to a zone situated closest to the origin (x_{max1} = 0.005), but it exhibits a maximum.

Our theoretical concept offers several interesting implementations. As a model hypothetical case, the coupling of the primary and secondary field forces of different intensities but the same nature can be used to demonstrate the relative importance of both fields to the resulting focusing phenomenon. The primary high-intensity field is assumed to generate only the transversal gradient in a binary carrier liquid. The secondary low-intensity field is applied simultaneously or subsequently to focus the components of the separated sample.

As a real case, the primary electric field can be applied, for example, to generate the density gradient in a suspension of charged colloidal particles and can be coupled with a secondary gravitational field to focus the components of the separated sample that practically are not affected by the primary electric field.

The computer simulation was performed to demonstrate the effect of the relative intensities of primary and secondary fields on the shape

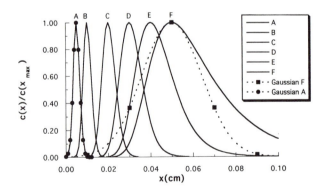

Figure 2. The concentration distribution of six focused zones (A–F) calculated by using the rigorous distribution function compared with Gaussian distributions at positions A and F. The input parameters are v_i = 5.236 $\times$ 10^{-13} cm^3, g_f = 981 cm/sec^2, $\phi_{m,ave}$ = 0.05, $\Delta\rho_m$ = 1.2 g/cm^3, w = 0.1 cm, $U_{m,e}/D_m$ = 100 cm^{-1}, x_{max1} = 0.005 cm, x_{max2} = 0.01 cm, x_{max3} = 0.02 cm, x_{max4} = 0.03 cm, x_{max5} = 0.04 cm, x_{max6} = 0.05 cm.

of the density gradient and on the shape and resolution of the focused zones (*11*). Typical results are shown in Figure 3. It clearly can be seen with both field intensities being equivalent to 30 G, very good resolution of all focused species is achieved. However, the focusing resolution decreases with decreasing intensities of the fields, and no focusing effect appears if both intensities are equivalent to 1 G (*11*). However, a good separation of the focused zones can be achieved by increasing the intensity of the primary field and by keeping the secondary field intensity constant, equivalent to the natural gravitation. The effect is also shown in Figure 3. When coupling the primary field equivalent to 1000 G with a secondary field equivalent to 1 G, very good resolution is obtained especially for the region between $x = 0$ and $x = 0.1$ mm, where the density gradient formed is steeper.

An important conclusion can be drawn from this computer simulation. The operational variables of focusing FFF experiments using a coupling of two effective field forces can be chosen in such a way that in many practically important cases, the natural gravitation can be exploited as a secondary field to focus the components of the separated sample. The apparatus and the experimental procedure to perform such a separation is very simple and inexpensive compared with classical centrifugation. The condition necessary to realize such a separation is that

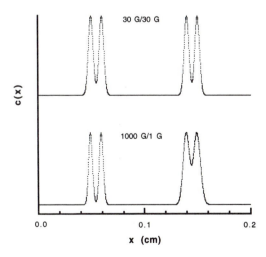

Figure 3. Computer simulation of static resolution of the four zones at different positions of the arbitrarily chosen focused species in a gradient forming carrier liquid by applying either both primary and secondary field intensities equivalent to 30 G or the coupling of the gradient generating primary field intensity equivalent to 1000 G with the secondary focusing field equivalent to 1 G.

the intensity of the primary field generating the density gradient is high enough.

As concerns a real case using the coupling of two fields of different nature, the previously mentioned electric and gravitational fields coupling can be considered. The charged particles of the density modifier colloid move in the electric field with the velocity $U_{m,e}$ proportional to their zeta potential. The conditions under which the same density gradients are formed due to either the centrifugal forces or the electric field action on the density modifier colloid particles can be estimated by using the ratio $|U_{m,g}| : |U_{m,e}|$:

$$\frac{|U_{m,g}|}{|U_{m,e}|} = \frac{2r_m^2 \, g_m \Delta \rho_m}{9 \epsilon \zeta E} \tag{10}$$

where ϵ is the permittivity and E is the electric field strength. The calculations have shown that the electric potential of about 5 mV applied across the separation channel of $w = 0.2$ mm should be equivalent to the primary centrifugal acceleration of 2000 G. Coupled with secondary natural gravitation of 1 G, this low potential can be applied for separation by isopycnic focusing in TLF or FFF.

Conversion of Resolution from Static to Dynamic Conditions. The separation by focusing proper can be exploited for analytical purposes, and the fractionation can be accomplished by the bulk flow of the carrier liquid in the direction of the x-coordinate under conditions that prevent the perturbation of the focused zones and thus the deterioration of the established separation. However, the bulk flow of the carrier liquid that represents the additional nonselective transport of all focused zones at identical velocities along the focusing axis does not contribute to the increase of the resolution. On the other hand, the bulk flow of the carrier liquid in the direction perpendicular to the direction of the focusing, such as in focusing FFF, can contribute to the increase of the resolution under optimal conditions. The resulting relationship for resolution under dynamic flow conditions is (9)

$$R_s = \sqrt{\frac{D_i L}{3\langle v \rangle}} \, \frac{w}{4m^2 \sigma^2} \, \frac{|1/(\Gamma_1 - \Gamma_1^2) - 1/(\Gamma_2 - \Gamma_2^2)|}{(1 - 2\Gamma_1)/(\Gamma_1 - \Gamma_1^2)^{3/2} + (1 - 2\Gamma_2)/(\Gamma_2 - \Gamma_2^2)^{3/2}} \tag{11}$$

where L is the length of the channel, $\langle v \rangle$ is the average linear velocity of the carrier liquid flow, parameter $m \approx 2$, σ is the standard deviation of the focused zone, and $\Gamma = x_{max,i}/w$. To evaluate the efficiency of the conversion from the static resolution, R_s, into dynamic resolution, R_s^*, under flow conditions, the conversion factor, f_c, was defined as (9)

$$f_c = R_s^*/R_s \qquad (12)$$

The calculations were carried out with some chosen input parameters by neglecting the differences in viscosities of the carrier liquid at positions of the focused zones (9). Few representative results are given in Table I. Without going into the details, it can be concluded that the static and dynamic conditions represent two basic configurations for the experimental arrangements of isoperichoric focusing either in thin-layer cell or in focusing FFF channel. The arrangement using the carrier liquid flow parallel with the direction of the focusing forces cannot be considered advantageous. As mentioned previously, the resolution cannot be improved because of the additional bulk flow transporting all focused zones at identical linear velocities. On the other hand, the experimental conditions in focusing FFF can be chosen in such a way that the resolution is higher compared with already established static resolution due to only focusing processes. The experimental conditions, the operational variables, and the parameters concerning the modifier as well as the separated species used for the model calculations do not represent the limits of the proposed methodology nor the extent of the potential applications. They have been chosen just to demonstrate clearly the flexibility of the experimental conditions that must be optimized for each particular application and this is why they do not correspond to actual experiments performed later. The isoperichoric focusing can be more efficient under static conditions in some instances but the resolution of the final separation can be improved by the flow provided that the experimental conditions are optimized.

Experimental Implementation. Various operational parameters affecting the formation of the density gradient generated by the electric field action on the charged colloidal silica particles and the focusing effect of larger particles were investigated experimentally under con-

Table I.   **Efficiency of the Resolution Conversion for Some Chosen Operational Parameters**

x_{max1}/x_{max2} (cm/cm)	R_s	f_c^a (0.1 cm/s)	f_c^a (0.01 cm/s)	R_s^b	f_c^b (0.01 cm/s)
0.001/0.002	3.053	3.217	10.17	3.053	3.217
0.003/0.004	2.762	2.169	6.860	2.762	2.169
0.005/0.006	2.499	1.685	5.329	2.499	1.685
0.007/0.008	2.262	1.414	4.472	2.262	1.414
0.009/0.010	2.046	1.291	4.084	2.046	1.291

^{a} g_f, 981,000 cm/s^2; v_i, 5.2×10^{-13} cm^3; w, 0.02 cm.
^{b} g_f, 981 cm/s^2; v_i, 5.2×10^{-10} cm^3; w, 0.02 cm.

ditions of static TLF and dynamic focusing FFF. To understand better the transport processes leading to steady-state gradient and to focusing, the properties and the behavior of the density gradient forming carrier liquid were studied in detail. The experimental results were compared with the theoretical predictions. The deviations of the experiments from the theoretical model are discussed in context of the potential application of the proposed concept not only for the separation purposes but also for the studies of interparticle interactions.

Simple apparatuses for TLF and focusing FFF experiments were constructed and described in detail in a previous article (13). Cylindrical TLF cell of 11 mm i.d. and of the variable height was used. The distance between the electrodes was equal to the height of the cell. The rectangular cross-section focusing FFF channel had the dimensions of 200 × 15 × 0.5 mm. Both TLF cell and focusing FFF channel are schematically represented in Figure 4. Percoll (Pharmacia Fine Chemicals AB, Uppsala, Sweden), which are colloidal silica particles coated with poly(vinyl pyrrolidone) and suspended in water, was used as density gradient to form carrier liquid. The original product was diluted with deionized water or with NaCl or sucrose solutions to obtain the suitable density carrier liquids. Density marker beads (Pharmacia Fine Chemicals

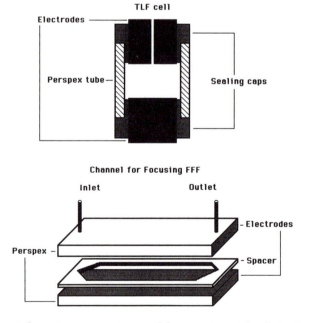

Figure 4.   Schematic representation of the experimental cell for thin-layer isopycnic focusing and of the channel for focusing FFF in coupled electric and gravitational fields.

AB), which are colored cross-linked dextran particles with defined buoyant densities, were used to determine the evolution and the final shape of the quasiequilibrium density gradients in TLF. Several samples of polystyrene (PS) latex microspheres (Duke Scientific Corporation, Palo Alto, CA), supplied by manufacturer as particle size standards, were used in focusing FFF experiments.

Formation of Steady-State Density Gradient in TLF Experiments. The formation of the density gradient by the effect of electric field on Percoll was evidenced by the appearance of distinct focused zones of density marker beads in TLF cell (*13*). A similar experiment was performed by using PS latex standard particles suspended in Percoll. Narrow steady-state focused zone was formed, thus indicating that focusing FFF of these particles should appear in the separation channel as well.

Although the true equilibrium density gradient is theoretically never reached in an ideal system, the near-equilibrium quasi-steady-state can be attained in real time according to the imposed operational parameters. The intention was to determine the conditions under which the focusing can be performed at or near the steady state with the given TLF experimental arrangement. Percoll diluted with deionized water and the density marker beads of different buoyant densities were used for this measurement. The evolution of the density gradient in TLF cell as a function of time was displayed by the positions of the focused zones of the individual density marker beads and is shown in Figure 5. The experimental steady state was reached within 500 min in a 10-mm TLF cell and within 90 min in a 5-mm cell (*13, 14*). This result agreed with the theoretical expectations and was encouraging for the focusing FFF

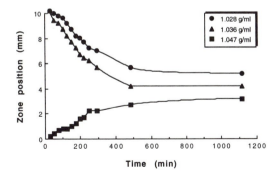

Figure 5. The evolution of the density gradient as a function of time in TLF cell displayed by the positions of the focused zones of three density marker beads suspended in Percoll diluted with water. (Reproduced with permission from reference 14. Copyright 1994 Springer-Verlag.)

experiment because the thickness of the separation channel is substantially reduced and, consequently, the time to reach quasi-steady state is proportionally shorter. The other experiments were performed under steady-state conditions determined for each particular case. Because of the preliminary character of the experiments, the quantitative comparison of the experimental results with the theoretical estimations was not relevant.

Another experiment was intended to compare the shape of the density gradient formed because of either centrifugation or coupling of electric and gravitational fields in TLF (13). The positions of the zones were almost identical for both experiments. A slightly steeper normalized density gradient was achieved by isopycnic centrifugation that might be due to higher average intensity and nonhomogeneity of the centrifugal field compared with the electric field or to higher height of the liquid column. No change in position of the steady-state zones was observed for several hours in TLF experiments when the electric field was turned off. This indicates that density marker beads sediment or float in stable density gradient due to gravitation without being affected by the electric field within the precision limits of the determination of the zone position.

The first approximation theoretical model did not explicitly take into account the interparticle forces. It was assumed that the ratio U_m: D_m is independent of concentration. When this ratio varies with the particle concentration because of the particle–particle interactions, especially in semidiluted and concentrated colloidal suspensions, the experimentally found and the theoretically calculated steady-state density distributions could be different. Such a difference can, in principle, be used to evaluate the contribution of the interparticle interactions to the resulting shape of the experimental density distribution (11, 19).

The study of the electric field strength effect on the shape of the density gradient formed in the TLF cell indicated an important difference compared with the first approximation theoretical model. A series of experimental data and the theoretically calculated curves are shown in Figure 6. The difference can be caused by the interactions between the colloidal particles of the binary density forming carrier liquid. Moreover, the electric field strength across the cell or channel thickness was estimated from the electric potential measured between the electrodes, but the electrochemical processes at both electrodes can contribute to this difference.

The addition of the electroneutral sucrose to the carrier liquid that is recommended for cell separations did not influence the formation of the density gradient in a measurable manner and can be used to extend the density range of the given carrier liquid.

The electrolyte added to the carrier liquid (or more exactly the ionic strength) influences the zeta potential of the colloidal particles and thus

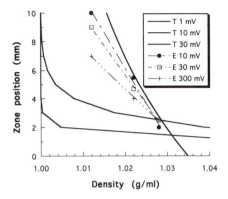

Figure 6. Theoretical (T) and experimental (E) shapes of the steady-state density gradients formed due to the different electric field strengths in Percoll diluted with water in TLF cell. Initial average density of Percoll was 1.024 g/mL. Experimental points correspond to the positions of the focused zones of density marker beads.

their mobility in the electric field and the thickness of the electrostatic double layer that modifies the interparticle forces depending on the distance between the particles. The formation of the density gradients in Percoll 0.15 M NaCl solution was not observed within the range of electric potentials from 10 to 1000 mV, independently of the average initial density of Percoll, but when decreasing the concentration of NaCl to 0.05 M in Percoll, a weak density gradient was formed.

Focusing FFF Experiments. The focusing FFF experiments were performed with PS latex samples having the particle diameter between 9.87 and 40.1 μm and by using diluted Percoll as the carrier liquid (*12–14*). Various electric potentials (0, 32, and 100 mV) across the channel were applied. The effect of the field was found to be different for each of these two PS latex samples. Relatively broad fractograms were obtained without electric field applied, the front eluting at 0.67 of the void volume, V_0, of the channel. The low diffusion coefficient of the large particles can explain this behavior because they do not have enough time to diffuse from the central fastest streamline to the lower velocity region during the time allowed for elution. The maximal velocity at the centerline of the rectangular cross-section channel is 1.5 times the average velocity of the carrier liquid considered as isoviscous in the first approximation; the corresponding elution volume is then 0.67 V_0. The elution volume of a low molecular weight solute having higher diffusion coefficient should be equal to V_0, which was roughly confirmed by the injection of the low molecular weight dye dissolved in carrier liquid.

The increase of the electric field strength resulted in a steeper density gradient, which caused the narrowing of the elution curves of PS latex of 9.87 mm while the position of the peak remained practically unchanged: PS latex particles formed the zone focused near the center line of the channel. The viscosity gradient across the channel, formed in consequence of the concentration gradient, results in perturbation of the parabolic flow velocity profile. However, the difference between isoviscous and nonisoviscous flow velocity profiles was found to be negligible and was not taken into account (12).

The behavior of the 40.1-μm PS latex sample was different. Although a single broad peak appeared without electric field applied, with the front eluting at $0.67 \times V_0$, two well-resolved peaks emerged when the density gradient was formed because of an active electric field. The elution volumes of $0.67 \times V_0$ of the first peaks corresponded to the unrelaxed part of the PS latex sample. The second peak, eluting at $0.86 \times V_0$, was relatively narrow and corresponded to the focused zone. The position of this focused peak did not change with the electric potential increasing from 32 to 100 mV, but the peak was narrower, more symmetrical, and its magnitude was higher than the unrelaxed peak. The presence of the "relaxation peak" indicates that the ratio of transversal migration (focusing) to longitudinal elution was not ideal. The channel dimensions and the experimental conditions should be studied in more detail so the phenomenon can be fully understood. When the flow rate was stopped after the injection for a relaxation period and restored until the elution was accomplished, a small change of the elution curve of 9.87-μm PS latex was observed compared with the experiment without stop flow. As concerns 40.1-μm PS latex, further substantial decrease of the height of relaxation peak was observed. When 9.87- and 40.1-μm PS latex samples were mixed and injected together, a good resolution was achieved, as shown in Figure 7. This separation indicates a small difference in densities or the difference in the ratio of the surface charge to the mass of PS latex samples. Although a good resolution of two samples was obtained by focusing FFF, the positions of the focused zones in the static TLF experiment were indistinguishable. In other words, this finding indicates at least qualitatively the expected superior resolution of the dynamic FFF separation over the static TLF experiment, especially when the focused zone position difference is very small. In terms of the density, the difference was about 0.0005–0.0010 g/mL.

The fractograms of PS 19.71-mm particles resulted again in a single relatively broad peak with the front eluting at 0.68 of V_0 (12). With the electric field applied, two peaks appeared. The main focused peak eluted at longer elution time when the intensity of the electric field was increased and a small unrelaxed peak appeared at 0.68 of V_0. The unrelaxed peak area decreased with increased intensity of the electric field.

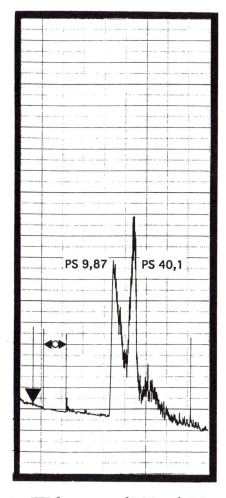

*Figure 7.   Focusing FFF fractogram of 9.87- and 40.1-μm PS latex micro-
spheres in diluted Percoll density gradient generated by the electric potential
of 32 mV and using the stop–flow procedure for relaxation. (Reproduced
with permission from reference 13. Copyright 1993 Springer-Verlag.)*

The relative elution volumes of the main peaks were 0.94 of V_0 and
1.27 of V_0 for two different electric potentials applied (100 and 150
mV, respectively). The relative elution volumes were inversely proportional to the relative velocities with respect to the average velocity of
the carrier liquid and corresponded to the positions of the focused zones
inside the channel.

These results could seem contradictory. However, the calculations
and the experimental results under static TLF conditions proved that
the focused zones can be displaced with the change of the electric

field to lower or to higher altitudes as well as they should not be displaced at all. The retentions in focusing FFF exhibit the same behavior.

Conclusions

It has been demonstrated, by using the model calculations, that various experimental arrangements can be applied to perform the isoperichoric focusing FFF and thin-layer isoperichoric focusing based on coupling of the gradient generated by a primary field with a secondary field action. The conversion of the resolution due to the transformation of the static focusing separation into the dynamic separation mechanism can be advantageous under carefully chosen conditions.

Experimental results demonstrated that the new separation concept is operating effectively but the design of the equipment, experimental conditions, and operational variables have to be optimized.

A presumed application of this concept is for analytical or continuous micropreparative separation and characterization of particles according to their densities. This use can be suggested for the separation of biological cells and of inorganic or synthetic polymer particles. However, more extensive investigation is needed to make exact conclusions.

Another possible usage is for the characterization of the gradient forming species and of their interactions with the primary field by determining the shape of the established gradient from the experimental retentions.

References

1. Janča, J. Makromol. Chem. Rapid Commun. 1982, 3, 887.
2. Janča, J. ISPAC4, 1997, Baltimore, MD, J. Appl. Polym. Sci., Appl. Polym. Symp. 1992, 51, 91.
3. Janča, J. Amer. Lab. 1992, 24, 15.
4. de Gennes, P. G. J. Chem. Phys. 1974, 60, 5030.
5. Giddings, J. C.; Li, S.; Williams, P. S.; Schimpf, M. E. Makromol. Chem., Rapid Commun. 1988, 9, 817.
6. Semyonov, S. N. Zh. Fiz. Khim. 1986, 60, 1231.
7. Janča, J.; Chmelik, J. Anal. Chem. 1984, 56, 2481.
8. Giddings, J. C. Separ. Sci. Technol. 1986, 21, 831.
9. Janča, J. Mikrochim. Acta 1994, 112, 197.
10. Kolin, A. In Electrofocusing and Isotachophoresis; Radola, B. J.; Graesslin, D., Eds.; de Gruyter: Berlin, Germany, 1977.
11. Janča, J.; Audebert, R. ISPAC5, 1999, Inugama, Japan, J. Appl. Polym. Sci., Appl. Polym. Symp. 1993, 52, 63.
12. Janča, J.; Audebert, R. J. Liq. Chromatogr. 1993, 16, 2211.
13. Janča, J.; Audebert, R. Mikrochim. Acta 1993, 111, 163.
14. Janča, J.; Audebert, R. Mikrochim. Acta 1994, 113, 299.

15. Meselson, M.; Stahl, F. W.; Vinograd, J. *Proc. Natl. Acad. Sci. U.S.A.* **1957,** *43,* 581.
16. Svensson, H. *Acta Chem. Scand.* **1961,** *15,* 325.
17. Giddings, J. C.; Dahlgren, K. *Sep. Sci.* **1971,** *6,* 345.
18. Schure, M. R.; Caldwell, K. D.; Giddings, J. C. *Anal. Chem.* **1986,** *58,* 1509.
19. Biben, T.; Hansen, J.-P.; Barrat, J.-L. *J. Chem. Phys.* **1993,** *98,* 7330.

RECEIVED for review January 6, 1994. ACCEPTED revised manuscript November 29, 1994.

Size-Exclusion Chromatography with Electrospray Mass Spectrometric Detection

William J. Simonsick, Jr.[1] and Laszlo Prokai[2]

[1] Marshall R&D Laboratory, DuPont, Philadelphia, PA 19146
[2] Center for Drug Discovery, College of Pharmacy, University of Florida, Gainesville, FL 32610

We interfaced a size-exclusion chromatograph (SEC) to a mass spectrometer operating in the electrospray mode of ionization. Stable electrospray conditions were obtained using a tetrahydrofuran mobile phase containing $\sim 10^{-5}$ M dissolved sodium salt, which affords pseudomolecular ions through cationization. Using the SEC with electrospray detection, we calibrated the SEC for an ethylene-oxide-based nonionic surfactant. The calibration standards were the surfactant oligomers themselves. The chemical composition distribution of acrylic macromonomers was profiled across the molecular weight distribution. A nonuniform chemical composition distribution was observed. Cross-linkers, additives and stabilizers, and coalescing solvents contained in a complex waterborne coating formulation were analyzed in a single experiment.

SIZE-EXCLUSION CHROMATOGRAPHY (SEC) is a popular technique used to obtain molecular weight distributions (MWD) on polymers and oligomers (*1, 2*). Traditionally, detection is accomplished by a differential refractive index (DRI) detector. Unfortunately, DRI provides little information about the polymer chemical composition. The use of mass spectrometry for detailed polymer analysis is becoming more established due to the proliferation of soft ionization techniques that afford intact oligomer or polymer ions with a minimal number of fragment ions (*3–11*). In addition to the MWD information, the specific chemical structure, including end groups and the distribution of monomer units in a co-polymer, is obtained. Furthermore, the data furnished by soft ionization

0065–2393/95/0247–0041$12.00/0

techniques are predictable. If a researcher hypothesizes about a given structure, the postulated structure has a molecular formula that in addition to the nominal mass yields an isotope envelope that can be compared with the theoretical isotope envelope based on the molecular formula. Because soft ionization affords molecular or pseudomolecular ions, such hypothesis can be evaluated.

Electrospray ionization mass spectrometry (ESIMS) is a soft ionization technique that has been widely applied in the biological arena. Owing to its extremely low detection limits and its ultrasoft ionization process, ESIMS has been the most successful method of coupling a condensed phase separation technique to a mass spectrometer. Under ESI conditions the sample liquid is introduced into a chamber with a hypodermic needle. An electrical potential difference (usually 2–4 kV) between the needle inlet and the cylindrical surrounding walls promotes ionization of the emerging liquid and disperses it into charged droplets. Solvent evaporation upon heat transfer from the ambient gas leads to the shrinking of the droplets and to the accumulation of excess surface charge. At some point the electric field becomes high enough (up to 10^9 V/cm) to desorb analyte ions. This widely accepted desorption model (12) relies on the existence of preformed ions in solution. In other words, the ions observed in the mass spectra were originally present as ionized molecules in solution.

Proteins and biopolymers are typically ionized through acid-base equilibria. Larger biopolymers acquire more than a single charge. A charge envelope results from the analysis of a single species. Because mass spectrometers separate ions based on the mass-to-charge ratio, increasing the number of charges can be used to extend the operable mass range. In fact, Fenn and Nohmi (13) have observed polyethylene glycols with molecular weights up to five million on a quadrupole mass analyzer with upper mass limits of 1500 Da.

Unfortunately, ESIMS has had limited applications on synthetic polymers (13, 14). Unlike biopolymers, many synthetic polymers have no acid or basic functional groups that can be used for ion formation. Moreover, each species of a unique molecular formula can give rise to a charge distribution envelope, thus further complicating the spectrum. To resolve multiple charge distribution envelopes, ultrahigh resolutions are required (15). Unfortunately, commercial quadrupoles generally yield only unit resolution throughout their mass range. Therefore, synthetic polymers that can typically contain a distribution of chain lengths and a variety of endgroups furnish quite a complicated mass spectrum, making interpretation nearly impossible.

The approach we used to circumvent the difficulties described previously are as follows. We used sodium cations dissolved in our mobile phase to facilitate ionization. To simplify the resulting ESI spectra, we reduced the number of components entering the ion source. Further-more, the multiple charged states were reduced by analyzing only small molecules that generally produce fewer charged states. Once we dem-onstrated that an electrospray signal was furnished in a tetrahydrofuran (THF) mobile phase, we evaluated the utility of SEC–ESIMS for SEC calibration, measurement of chemical composition distribution in co-polymers, and complex mixture analysis.

Experimental Details

SEC was carried out using a three-column set of 10^3, 500, and 100 Å 30 cm × 7.8 mm i.d. Ultrastyragel columns (Waters, Milford, MA). The acrylic macromonomers were analyzed using a two-column set of 500- and 100-Å Ultrastyragel columns. The THF mobile phase was delivered by a Spectro-flow 4000 solvent delivery system (Kratos Analytical, Manchester, UK) at 1.0-mL/min flow rate. THF is a flammable solvent and proper care should be exercised when using large volumes. All samples under study were dis-solved in the mobile phase (~0.5% wt/vol) before analysis. The sample solutions were injected using a Rheodyne 7125 valve equipped with a 100-μL loop (Cotai, CA). Effluent splitting was achieved with a T-junction (Valco) that supplied only ~8–10 μL min/flow to the mass spectrometer through a 25-cm-long fused silica capillary (25 μm i.d.). A Spectroflow 757 absor-bance detector (Kratos Analytical) operated at 254 nm was used for deter-mining MWD data by SEC. The polystyrene calibrants used were molecular weights of 580, 2450, 5050, 11,600, and 22,000 Da with polydispersities ranging from 1.03 to 1.09. A block diagram of the SEC–ESIMS is seen in Figure 1.

A Vestec 200ES instrument (Vestec Corp., Houston, TX) was used to obtain the ESIMS (16). The spray was generated from the solvent entering

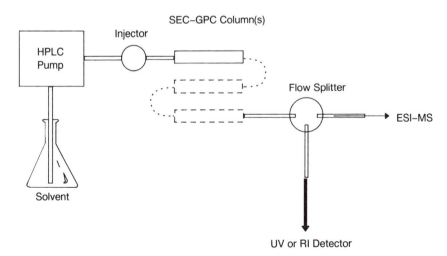

Figure 1.   Block diagram of the SEC–ESIMS instrumental setup used in all studies.

the ion source through a 0.005 i.d. × 0.010 o.d. flat-tipped hypodermic needle held at 3.0 kV potential. Preformed ions were obtained by dissolving ~5 × 10^{-5} M sodium iodide in the THF mobile phase. When the needle tip to orifice distance was ~10 mm, the spray current was in the range of 60–100 μA at 5–10 μL/min flow rate. The source block was heated to 250 °C, and the spray chamber temperature was estimated at 55–60 °C. A Vector/One data system (Teknivent, St. Louis, MO) was used to control the quadrupole analyzer (m/z 200–2000 at 3 ms/Da scan speed). Selected-ion chromatograms were reconstructed from full-scan data, and elution volumes ($V_E = t_E$ × flow rate) were determined by adjusting the time (t_E) versus intensity data to the theoretical (Gaussian) elution (17) profile using a nonlinear curve-fitting program running on an IBM-PC/AT-compatible computer (MINSQ, Micromath Scientific Software, Salt Lake City, UT).

The 80/20 (wt/wt) methyl methacrylate (MMA) n-butyl acrylate (BA) macromonomer was prepared in the following manner. To a 3000-mL flask 440.1 g MMA, 200.0 g BA, and 150.0 g methyl ethyl ketone (MEK) were added. The mixture was stirred and heated to reflux under a nitrogen blanket. After a 10-min hold, 30.0 g MEK, 0.140 g Vazo-67, and 0.050 g Co(dimethylglyoxime-BF$_2$)$_2$ were added to the flask. After a 5-min hold, 359.9 g MMA, 200.0 g MEK, and 1.90 g Vazo-67 were added over a 3.5-h period. The mixture was held 1 h at reflux after the feed. Subsequently, 150.0 g MEK and 1.00 g Vazo-52 were feed over an hour. The mixture was held for 1 h at reflux. The mixture was then allowed to cool to room temperature. A more detailed procedure and the Co(dimethylglyoxime-BF$_2$)$_2$ synthesis are given in reference 18.

Results and Discussion

Cationization has been the preferred technique for producing gaseous ions from synthetic oligomers and polymers by desorption ionization (3, 4, 6, 19). We have relied on this approach upon considering the coupling of ESI to SEC. A small amount, ~10^{-5} M, sodium iodide dissolved in the THF mobile phase does not impair the chromatography and affords meaningful ESI mass spectra as singly charged ions are seen as M[Na$^+$], doubly charged as M[2Na$^+$], and triply charged as M[3Na$^+$]. No ESIMS signal was observed without addition of a soluble salt that provided the source of cations.

To simplify the complexity of the resulting ESI spectrum, we chose to reduce the number of components entering the ESI source. We selected SEC because the mode of separation is well understood, predictable, and performed on a routine basis in our laboratory. To reduce the breadth of the charge envelopes, we chose to examine exclusively lower molecular weight materials, typically <3000 Da. This molecular weight regime encompasses many of the components contained in today's high-performance coatings (19).

We have previously reported on the coupling of an SEC to a mass spectrometer operated in the electrospray mode of ionization and its application to the molecular weight characterization of octylphenoxy-poly(ethoxy)ethanol oligomers (20). The analysis of nonionic surfactants

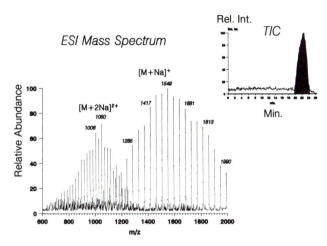

Figure 2. Electrospray ionization mass spectrum of octylphenoxy-poly(ethoxy)ethanol. Inset is the total ion chromatogram. Conditions are given in Experimental Details.

has also been accomplished by other condensed-phase separation techniques with mass spectrometric detection (*21, 22*). The following discussion serves as an introductory example of the data and its interpretation. For a more detailed discussion on the coupling of SEC to ESIMS and its application to octylphenoxypoly(ethoxy)ethanol, consult reference 20.

Figure 2 shows the summed electrospray mass spectrum of an octylphenoxypoly(ethoxy)ethanol (*see* 1). The total ion chromatogram is seen in the inset. The shaded region of the total ion chromatogram was summed to yield the spectrum. Two envelopes are present, due to singly and doubly charged species. Triply charged ions are also present but are slightly masked by the doubly charged envelope. We did not observe any other chemical species in this surfactant other than 1 nor did we target our analysis toward such materials. Investigators have previously reported low-level impurities in similar mixtures (*23*).

The determination of charge state is easily accomplished by examining the repeat group. The molecular weight of the ethylene oxide unit ($-CH_2CH_2O-$) is 44 Da, hence the spacing in the region above 1200

$$C_8H_{17} - \langle \text{phenyl} \rangle - (OCH_2CH_2)_n - OH \; + \; Na^+$$

| 189 Da | 44n Da | 40 Da |

1

Da. Close examination of the region from 600 to 1200 Da shows a 22-Da repeat unit due to the doubly charged oligomers. The base peak seen at 1549 Da is due to the sodiated $n = 30$ oligomer. The octylphenyl [C_8H_{17}–C_6H_4–] moiety contributes 189 Da. The 30 ethylene oxide groups add 1320 Da, whereas the terminal hydroxyl group and sodium cation contribute 40 Da; hence, the peak at [189 Da + 1320 Da + 40 Da] = 1549 Da. The equation describing the distribution of singly charged oligomers is

$$M[Na]^+ = (229 + 44n) \text{ Da} \qquad (1)$$

The doubly charged species follow the equation

$$M[2Na]^{2+} = (126 + 22n) \text{ Da} \qquad (2)$$

Likewise, the triply charged species are described by the equation

$$M[3Na]^{3+} = (91.67 + 14.67n) \text{ Da} \qquad (3)$$

The SEC elution behavior of any oligomer can be profiled by plotting the selected-ion chromatograms that correspond to the ions defined by equations 1–3. For example, the singly charged $n = 20$ oligomer furnishes a singly charged ion at 1109 Da (*see* eq 1). The doubly charged $n = 35$ oligomer yields ion at 896 Da (*see* eq 2). The triply charged $n = 50$ oligomer affords an ion at ~825 Da as defined by equation 3. Figure 3 displays the UV chromatogram ($\lambda = 254$ nm), the selected-ion plots of the singly charged $n = 20$ oligomer (1109 Da), the selected-ion plot of the doubly charged $n = 35$ oligomer (896 Da), and the triply charged $n = 50$ oligomer (825 Da) for the SEC analysis of octylphenoxypoly(ethoxy)ethanol (1). The fitted curves were generated using the nonlinear curve-fitting program described in Experimental Details.

As expected, the higher molecular weight $n = 50$ oligomer with a larger hydrodynamic volume elutes before (19 min) the smaller $n = 35$ oligomer (20 min) and $n = 20$ oligomer (22 min). Such data can be used to calibrate the SEC and unrelated calibrants such as narrow molecular weight polystyrenes can be avoided (*20*).

MWD information can be computed using the averaged mass spectrum presented in Figure 2. The doubly charged envelope would be used because a portion of the singly charged envelope exceeds the upper mass limit of our system. However, the reliability of this approach is poor due to instrumental parameters that may provide a nonuniform response with molecular weight. Moreover, the electrospray response is not uniform with increasing molecular weight. For example, the molecular weight average computed from the singly charged envelope is lower than that calculated from the doubly charged envelope. We speculate that the multiple charging becomes predominant and attenuates

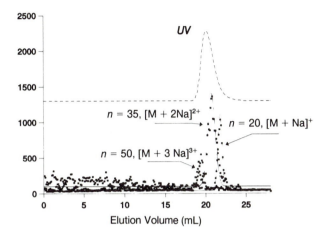

Figure 3. UV chromatogram (λ = 254 nm) and selected-ion traces for octylphenoxypoly(ethoxy)ethanol. The triply charged n = 50 oligomer selected-ion trace was obtained by summing 824–826 Da through the duration of the chromatogram. The doubly charged n = 35 oligomer selected-ion trace was furnished by summing 895–897 Da. The singly charged n = 20 oligomer selected-ion trace was obtained by summing 1108–1110 Da.

the relative abundance of the singly charged species proportionately to the increase in molecular weight.

We recommend that selected ion profiles be used for calibration. A full scan acquisition is collected and the selected-ion profiles of the individual oligomers used for calibration. Hence, the SEC is calibrated for octylphenoxypoly(ethoxy)ethanol using the octylphenoxypoly(ethoxy)ethanol oligomers. Figure 4 presents the calibration curve obtained from the selected-ion plots of individual oligomers. Also plotted is the calibration curve obtained using narrow molecular weight polystyrenes. Notice the large discrepancy at lower molecular weights (<2500 Da). This is the molecular weight region where polymers do not yet exhibit random coil behavior that could account for the discrepancy in calibration between polystyrene and octylphenoxypoly(ethoxy)ethanol. However, other factors such as solvent quality and column adsorption could also give rise to differences in retention behavior. Work is ongoing to explain the calibration differences seen at low molecular weights. In addition to providing accurate calibration, SEC–ESIMS is useful for determining how well the SEC is working for a given separation. For example, unwanted adsorption behavior can easily be monitored because compounds will elute at much later retention volumes and generally the peaks are significantly broadened. The suspect species can be monitored by profiling the sodiated ions associated with this component.

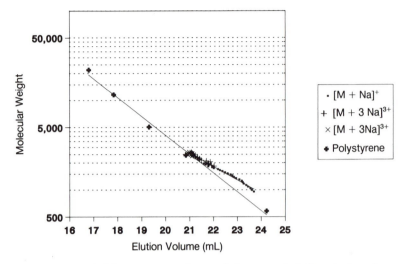

Figure 4.   Calibration curve for octylphenoxypoly(ethoxy)ethanol.

Acrylic Macromonomers.   Acrylic–methacrylic macromonomers prepared by catalytic chain transfer using cobalt(II) chelates afford polymer chains, each with an olefinic endgroup (*18*, *24*). Such macromonomers can be polymerized or copolymerized to produce graft polymers that are useful in coatings, fibers, films, and composite materials applications (*24*). Moreover, one is able to synthesize macromonomers containing several alkylmethacrylates, alkylacrylates, and styrene (*18*).

Using SEC–ESIMS we studied the products of a macromonomer synthesis in which MMA (**2**) and BA (**3**) MS were loaded in an 80:20 (wt/wt) weight ratio. The details of the synthesis are reported in Experimental Details. The SEC–ESIMS data will allow us to profile the chemical composition distribution across the MWD. From these data we should be able to measure the relative efficiency of our chain-transfer agent for methacrylates versus acrylates.

Using soft ionization via ESI, all oligomeric chains can be predictively examined. For example, MMA weighs 100 Da; therefore, oligomeric chains that contain only MMA will follow equation 4, where *n* is the number of MMA units. The 23-Da offset is due to the sodium cation.

MMA = 100 Da   BA = 128 Da

2   **3**

$$MMA_nBA_0 [Na^+] = (23 + 100n) \text{ Da} \tag{4}$$

All chains are vinyl-terminated and initiated via a hydride $(H \cdot)$. Sodium provides the ionization source. The MMA trimer, for example, is expected to produce an ion at 323 Da. If a BA, which has a molecular weight of 128 Da, is substituted into the MMA trimer chain, a net gain of 28 Da is expected, thus affording a 351-Da ion. Equation 5 describes the MMA chains possessing one BA, where n is the number of MMA groups.

$$MMA_nBA_1 [Na^+] = (151 + 100n) \text{ Da} \tag{5}$$

Addition of another BA in the MMA chain produces a gain of 28 Da, hence equation 6:

$$MMA_nBA_2 [Na^+] = (279 + 100n) \text{ Da} \tag{6}$$

Another BA gives rise to Equation 7:

$$MMA_nBA_3 [Na^+] = (307 + 100n) \text{ Da} \tag{7}$$

We did not observe any multiple charging for the macromonomers by using the electrospray parameters reported in Experimental Details. Nor did we observe any significant ions corresponding to oligomeric chains possessing greater than three BA groups. Unfortunately, the mass range of our mass spectrometer was limited to only 2000 Da; therefore, any ions in excess of 2000 Da were not recorded.

Initially, we wanted to know whether the addition of a BA into a MMA chain would change its hydrodynamic volume in THF; thereby giving rise to independent calibration curves for MMA_n homopolymer, MMA_nBA_1 copolymer, MMA_nBA_2 copolymer, and MMA_nBA_3 copolymer. Figure 5 presents the calibration curve for all oligomers. The correlation coefficient (r^2) is >0.99 for $n = 24$. However, at the 95% confidence level we found no significant differences among the calibration curves for the individual oligomers with 0, 1, 2, and 3 BA groups. This is not surprising because the inherent viscosity molecular weight relationships of the homopolymers are not significantly different, indicating similar chain dimensions.

We wanted to monitor the chemical composition distribution across the MWD. This can be accomplished by summing all the selected-ion plots for the MMA_nBA_0, MMA_nBA_1, MMA_nBA_2, and MMA_nBA_3, respectively. We used a 2-Da window to obtain the selected-ion plots. For example, the selected-ion plot for the MMA_2BA trimer (351 Da; eq

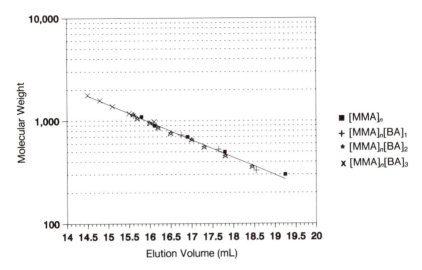

Figure 5.   Calibration curve for 80/20 MMA–BA macromonomer.

7) was obtained by integrating the ion current over the window of 350–352 Da.

Figure 6 plots each of the species previously discussed grouped according to the number of BA monomers in the chain. The dashed line overlayed in each block represents the total ion current. Each individual oligomer was profiled and grouped with other oligomers having the same number of BA units in the polymer chain. The individual selected-ion profiles were summed to yield the overall distribution of a given number of BA monomer units in the polymer chain.

In the specific example of the 80–20 MMA–BA, we see that the chemical composition is not uniform across the MWD. Specifically, we see the MMA_nBA_0 oligomer concentration maximizes late in the chromatogram. This region of the chromatogram corresponds to lower molecular weight. As the BA content in the macromonomer is raised, the distribution becomes nonuniform with BA-rich oligomers skewed toward higher molecular weights and shorter SEC retention volumes. These findings suggest that the chain-transfer efficiency of the cobalt complex is greater toward methacrylates versus acrylates.

The synthesis of graft copolymers from the MMA–BA macromonomer yields a comb polymer whereby the teeth of the comb are the macromonomer units. Therefore, the short teeth of the comb polymer will be richer in MMA than the sample loading (>80% wt/wt) and, conversely, the long teeth will be richer in BA (>20% wt/wt).

SEC–ESIMS provides a tool to monitor the copolymerizations of monomers in the presence of chain-transfer catalysts. Hence, we can

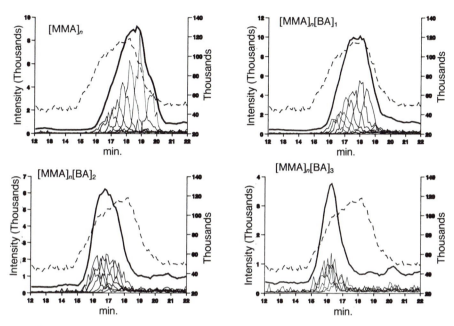

Figure 6.   Selected-ion plots of MMA$_n$, MMA$_n$BA$_1$, MMA$_n$BA$_2$, and MMA$_n$BA$_3$ oligomers.

measure the relative rates toward chain transfer of the two monomers in a copolymerization. It has been suggested that novel macromonomers and functionalized polymers could be synthesized using monomers with varying degrees toward catalytic chain transfer (25). SEC–ESIMS provides information about the chain-transfer efficiencies toward different monomers. The only prerequisite is that the monomers differ in their respective molecular weights. If the molecular weights of the monomers differ, soft ionization mass spectrometry should be adequate for determining reactivity ratios. Indeed, this has been accomplished by Montaudo and Montaudo (26), who used laser desorption Fourier transform mass spectrometry data published by Wilkins and co-workers (3) to make some qualitative predictions about the copolymerization. However, this approach assumes no fragmentation and accurate response with increasing molecular weight, which is likely not the case. SEC–ESIMS will allow the investigator to rapidly assess whether fragmentation has occurred. Moreover, SEC–ESIMS provides another detector response that can be compared with UV or RI. Subsequently, we can evaluate the ESIMS response for oligomers as a function of increasing molecular weight.

Complex Mixture Analysis.   High-performance organic coatings are extremely complex formulations. Specifically, waterborne automotive clearcoats are composed of a binder system and cross-linker in an appropriate solvent formulation. In addition, stabilization packages (antioxidants and photostabilizers) are added to lengthen the usable life of the coating. Each of the previous species contain a variety of chemical functionalities encompassing a wide range of molecular weights. Any separation method that can simplify the number of components for spectroscopic characterization is welcomed by the coatings chemist. SEC provides a predictable separation method to reduce the complexity of a formulation. Unfortunately, DRI detection yields no information about chemical composition, even though peaks may be fully resolved. Without the use of authentic materials, positive identification based on SEC with DRI detection is difficult. SEC–ESIMS affords exclusively molecular ion information that can be translated into plausible molecular formulae from which likely structures can be proposed.

Figure 7 displays the total ion chromatogram of a waterborne clearcoat formulation along with two averaged electrospray mass spectra. The first electrospray spectrum was obtained by summing the region

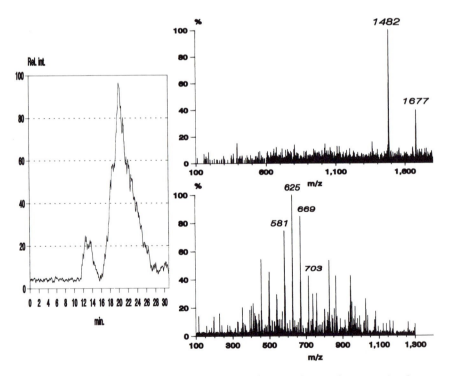

Figure 7. *(Left) Total ion chromatogram of a waterborne clearcoat. (Right, Top) summed electrospray mass spectrum (19.68–19.98 min); (right, bottom) summed electrospray mass spectrum (22.48–23.38 min).*

from 19.68 to 19.98 min. Two significant ions are present at 1482 and 1677 Da. We suspected these ions resulted from the stabilizer package as they fall into the molecular weight range of many commercial stabilizers. Because only pseudomolecular ions are formed under ESIMS conditions, we can compute the molecular weight and search the molecular weight using the Formula Weight Index of the Registry File of Chemical Abstracts Service (*27, 28*). A literature search of stabilizers and a formula weight of 1654 [1677 − 23 Da (sodium)] yielded a single compound, MIXXIM HALS 63. This compound with a molecular formula is a hindered amine light stabilizer with a molecular formula of $C_{91}H_{158}N_6O_{20}$ and is a likely candidate for the ion found in the waterborne clearcoat. Unfortunately, the resolution of our mass spectrometer was not adequate to compare the experimental isotope envelope with theory. A single ion with a molecular weight that agrees with the molecular weight of a hindered amine light stabilizer found in *Chemical Abstracts* is not unambiguous proof of the presence of this material. However, an authentic sample can be purchased and checked for its presence in the complex formulation. We were not successful using a similar approach for the identification of the 1482-Da ion.

The second electrospray spectrum obtained by summing the region from 22.48 to 23.38 min has several species present. Analysis by ^{13}C NMR showed the presence of a methoxymethylmelamine. ESIMS provides detailed characterization about the functionality; the molar amounts of formaldehyde; and distribution of monomers, dimers, trimers, and so on. These parameters affect the final properties of melamine cross-linked films. We find in this SEC window a large abundance of methoxymethylmelamine dimers. For example, the base peak at 625 Da we attribute to the dimer condensate seen in **4**. The 44-Da spacing giving rise to the 581-, 669-, and 713-Da ions is due to replacement of a methylene methoxy group [$-CH_2-O-CH_3$; 45 Da] by a proton [H; 1 Da], hence 45 − 1 Da = 44 Da. Such high-imino high-solids melamine-type resins are recommended for use in waterborne systems (*29*). Numerous ions were observed, many of which could be attributed to specific structures. For example, we found a large 185-Da ion accounting for

$[C_{22}H_{42}N_{12}O_8]\,Na^+ = 625\;Da$

4

the majority of ion signal at retention times over 25 min. This species is due to sodiated butylcarbitol (molecular weight 162 Da), the coalescing solvent. We do not recommend SEC–ESIMS for solvent analysis; however, the information is available. As previously discussed, the SEC–ESIMS also allows one to evaluate whether any decomposition or fragmentation has occurred. Moreover, in the direct mass spectrometric analysis of complex mixtures, the thermal decomposition of large polymers can sometimes complicate and obscure lower molecular weight ions. This complication is not a problem using SEC–ESIMS because the high molecular weight materials are separated from the low molecular weight species by the SEC; therefore, decomposition products are readily detected.

ESI has been used to extend the molecular weight range on commercial mass spectrometers by using the multiple charging. We hoped such an approach could be used for polymer analysis. Unfortunately, the relatively low resolution of our current system did not provide any specific information on high polymers (>10,000 Da). For example, the peak seen eluting from 12 to 16 min furnished an ESI spectrum that was a continuum of peaks that gave no additional information than the DRI detector provided.

Summary

SEC–ESIMS is a valuable tool for polymer characterization. Compounds are separated based on their hydrodynamic size in solution, but upon detection, an absolute molecular weight is also furnished. Only 1% of the SEC effluent is required for ESIMS analysis, thereby accommodating the popular SEC detectors. SEC–ESIMS provides an attractive solution to the calibration of SEC without the use of external calibrants. Chemical composition distribution information on copolymers is easily afforded provided the individual monomers differ in molecular weight. The successively acquired mass spectra contain narrow fractions of the overall distribution that simplifies the analysis of complex formulations. Unfortunately, we have not been able to provide structured details on materials beyond 5000 Da due to the low resolution of the quadrupole mass spectrometer. Nevertheless, SEC–ESIMS is an exciting hyphenated techniques for polymer characterization.

Acknowledgments

The Automotive Sector of DuPont is thanked for the permission to publish this work. Nicholas Bodor, Director of the University of Florida Center for Drug Discovery, provided much of the instrumentation. Andrew H. Janowicz synthesized the macromonomers. Audrey Lockton is gratefully acknowledged for her clerical assistance in the preparation of this manuscript.

References

1. Yau, W. W.; Kirkland, J. J.; Bly, D. D. *Modern Size-Exclusion Liquid Chromatography;* John Wiley & Sons: New York, 1979.
2. Bidlingmeyer, B. A.; Warren, V. F., Jr. *LC-GC* **1988**, *6(9)*, 780–786.
3. Nuwaysir, L. M.; Wilkins, C. L.; Simonsick, W. J., Jr. *J. Am. Soc. Mass. Spectrom.* **1990**, *1*, 66–71.
4. Prokai, L.; Simonsick, W. J., Jr. *Macromolecules* **1992**, *25*, 6532–6539.
5. Vouros, P.; Wronka, J. W. In *Modern Methods of Polymer Characterization;* Barth, H. G., Mays, J. W., Eds.; John Wiley & Sons: New York, 1991; pp 495–555.
6. Schulten, H. R.; Lattimer, R. P. *Mass Spectrom. Rev.* **1984**, *3*, 231–315.
7. Lattimer, R. P.; Sculten, H. R. *Anal. Chem.* **1989**, *61*, 1201A–1215A.
8. Prokai, L. *Field Desorption Mass Spectrometry;* Marcel Dekker: New York, 1989; pp 254–262.
9. Benninghoven, A.; Rudenauer, F. G.; Werner, H. W. *Secondary Ion Mass Spectrometry: Basic Concepts, Instrumental Aspects, Applications and Trends;* John Wiley & Sons: New York, 1987; pp 739–744.
10. Biemond, M. E. F.; Geurts, F. A. J.; Mahy, J. W. G. In *Secondary Ion Mass Spectrometry (SIMS VIII);* Benninghoven, A., Ed.; Springer-Verlag: Berlin, Germany, 1989; pp 811–814.
11. Montaudo, G.; Scamporrino, E.; Vitalini, D. In *Appl. Polym. Anal. Charact.;* Mitchell, John, Jr., Ed.; Hanser: Munich, Germany, 1992; Vol. 2, pp 79–102.
12. Iribarn, J. V.; Thomson, B. A. *J. Chem. Phys.* **1976**, *64*, 2287–2294.
13. Fenn, J. B.; Nohmi, T. *J. Am. Chem. Soc.* **1992**, *114*, 3241–3246.
14. Kallos, G. J.; Tomalia, D. A.; Hedstrand, D. M.; Lewis, S.; Zhou, J. *Rapid Comm. Mass Spectrom.* **1991**, *5*, 383–386.
15. Beu, S. C.; Senko, M. W.; Quinn, J. P.; Wampler, F. W., III; McLafferty, F. W. *J. Am. Soc. Mass Spectrom.* **1993**, *4*, 557–565.
16. Allen, M. H.; Vestal, M. L. *J. Am. Soc. Mass Spectrom.* **1992**, *3*, 18–26.
17. Giddings, J. C. *Dynamics of Chromatography, Part I, Principles and Theory;* Marcel Dekker: New York, 1965; p 13.
18. Janowicz, A. H.; Melby, L. R. U.S. Patent 4 680 352, 1987.
19. Simonsick, W. J., Jr. *Prog. Org. Coat.* **1992**, *20*, 411–423.
20. Prokai, L.; Simonsick, W. J., Jr. *Rapid Comm. Mass Spectrom.* **1993**, *7*, 853–856.
21. Lee, M. L.; Markides, K. E. *Analytical Supercritical Fluid Chromatography and Extraction;* Chromatography Conferences: Provo, UT, 1990; pp 475–479.
22. Yergey, A. L.; Edmonds, C. G.; Lewis, I. A. S.; Vestal, M. L. *Liquid Chromatography/Mass Spectrometry Techniques and Applications;* Plenum: New York, 1990.
23. Schultz, G. A.; Alexander, J. N. In *Proceedings of the 42nd American Society for Mass Spectrometry;* Conference on Mass Spectrometry and Allied Topics; American Society for Mass Spectrometry: Santa Fe, NM, 1994; p 174.
24. Janowicz, A. H. U.S. Patent 4 722 984, 1988.
25. Greuel, M. P.; Harwood, H. J. *Polym. Prepr., Am. Chem. Soc. Dev. Polym. Chem.* **1991**, *32(1)*, 545–546.

26. Montaudo, M. S.; Montaudo, G. *Macromolecules* **1992**, *25*, 4264–4280.
27. Leiter, D. P.; Morgan, H. L.; Stobaugh, R. E. *J. Chem. Document* **1965**, *5*, 238.
28. Dittmar, P. G.; Stobaugh, R. E.; Watson, C. E. *J. Chem. Inform. Comp. Sci.* **1976**, *16*, 111.
29. *Resimene: Amino Crosslinker Resins for Surface Coatings;* Monsanto Technical Bulletin #6515B; Monsanto Corp., MTS Department: Springfield, MA, 1986.

RECEIVED for review January 6, 1994. ACCEPTED revised manuscript November 18, 1994.

LIGHT SCATTERING AND VISCOMETRY: MULTIDETECTION CALIBRATION AND APPLICATION

5

Concerns Regarding the Practice of Multiple Detector Size-Exclusion Chromatography

Christian Jackson and Howard G. Barth

Corporate Center for Analytical Sciences, DuPont, Experimental Station, Wilmington, DE 19880–0228

The use of multiple detectors with size-exclusion chromatography (SEC) can greatly increase the information content available from a typical SEC analysis. This multidetector approach permits more accurate measurement of polymer properties than conventional SEC. The additional information, however, is obtained at the expense of an increase in the complexity of the instrumentation and data handling. In particular, a number of concerns arise in data acquisition and processing that are not present in conventional SEC. Some of these difficulties are outlined, and possible solutions are discussed.

MEASUREMENT OF THE SIZE DISTRIBUTION of molecules present in polymeric materials is essential to understanding polymer physical properties; in addition, the size distribution contains information about the polymerization process. The development of size-exclusion chromatography (SEC) made the rapid measurement of relative molecular weight distributions (MWD) possible with high precision (*1, 2*). It has become possible to combine a number of classical polymer characterization techniques with SEC and thus to measure polymer properties for each molecular weight fraction of the distribution. Additional detectors used in combination with SEC can greatly increase the information content available from a typical analysis. These detectors can include, for example, a molecular-weight-sensitive detector, such as a continuous viscometer or light-scattering detector, or a spectroscopic detector, such as a UV spectrophotometer. This multidetector approach

0065–2393/95/0247–0059$12.00/0
© 1995 American Chemical Society

permits more accurate measurement of polymer properties than conventional SEC (*3–10*). Absolute MWD can be measured using light scattering or viscometry combined with universal calibration. Compositional drift over the MWD of a polymer can be measured using a UV spectrophotometer and a differential refractive index detector. The increase in the available information also expands the complexity of data analysis. We discuss some of the concerns regarding data analysis that arise in multidetector SEC.

Experimental Details

The chromatograph consisted of a model 590 pump, a model 710B automatic injection module, and a model 410 differential refractometer (Waters Associates, Milford, MA). The column set consisted of two PLgel 5-μm mixed bed (MIXED-C) 300 × 7.5-mm columns (Polymer Laboratories, Amherst, MA). The viscometer was a model 110 (Viscotek Corporation, Houston, TX) and the light-scattering detector was a model F (Wyatt Technology Corporation, Santa Barbara, CA). The polymer samples were narrow and broad MWD polystyrenes (PS) (Polymer Laboratories, and American Polymer Standards, Mentor, OH). The mobile phase was high-performance liquid chromatography grade tetrahydrofuran (Aldrich Chemical Company, Milwaukee, WI). A 0.5-μm filter was placed between the pump and the autoinjector (Millipore, Milford, MA).

Discussion

Detector Configuration. The detectors can be arranged in series or in parallel after the SEC columns. For three or more detectors, combinations of series and parallel connections are possible. The advantage of parallel configurations is that they avoid additional peak broadening caused by the eluting sample passing through a number of detector cells. Series configurations provide greater control over flow rate fluctuations because of back pressure variations among detectors.

To determine the relative importance of these different effects, the peak widths for narrow MWD PS standards were measured in three different configurations of the instrument. The three configurations studied are shown schematically in Figure 1. The instruments were connected using the shortest possible lengths of 0.25-mm diameter stainless steel tubing. The refractometer and viscometer tracings from equal concentrations of PS with nominal molecular weights of 600,000 and 200,000 g/mol are shown in Figure 2.

The sharpest peaks were obtained with the viscometer preceding the refractometer in a series configuration (Figure 2A). The number of theoretical plates (*N*) for the respective 600 and 200 K g/mol PS standards was 1000 and 3000 for the viscometer and 2000 and 4300 for the refractometer. Because of the large dead volume in the refractom-

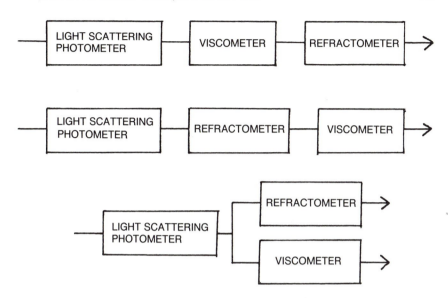

Figure 1.   *The three detector configurations studied.*

eter, it could not be placed before the viscometer (Figure 2B). In this case, N for the respective 600 and 200 K g/mol PS standards was 600 and 300 for the viscometer and 2000 and 4200 for the refractometer. In the parallel configuration the viscometer peaks are slightly broader, but the refractometer peaks are similar to the first series configuration (Figure 2C). This may be due to flow rate fluctuations or mixing occurring in the T-junctions used to split the flow. In this configuration, N for the respective 600 and 200K g/mol PS standards was 700 and 1100

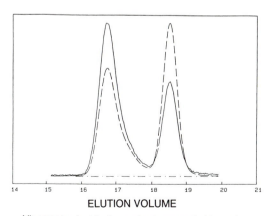

ELUTION VOLUME

Viscometer (—) before refractometer (- -) in series.

Figure 2A.   *Detector output tracings from the viscometer (———) placed before the refractometer (- – –) in series.*

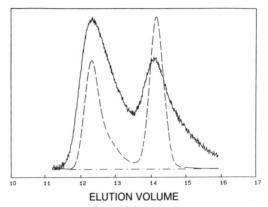

ELUTION VOLUME

Refractometer (- -) before viscometer (—) in series.

Figure 2B.   Detector output tracings from the refractometer (– – –) placed before the viscometer (——) in series.

for the viscometer and 1500 and 3000 for the refractometer. These results suggest that a series configuration produces the best data; however, band broadening and peak distortion are clearly very dependent on the measurement cell design and the volume of interconnecting tubing of the instrument used. Although the data are not shown here, a similar trend was observed for the light-scattering–refractometer combination.

Interdetector Volume.   Because the detectors are placed at different physical positions in the elution stream, the detector signals are

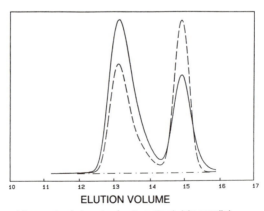

ELUTION VOLUME

Viscometer (–) and refractometer (--) in parallel.

Figure 2C.   Detector output tracings from the refractometer (– – –) and the viscometer (——) in parallel.

initially misaligned. The volume between the detectors must be known and compensated before data analysis (*11–13*). Any errors in determining this volume can produce significant errors in the measurement of sample molecular weight polydispersity (*14, 15*). The light-scattering and viscosity peaks appear at lower elution volumes than the refractometer peak because of the detectors increased sensitivity to higher molecular weight species. The magnitude of this shift depends on the polydispersity of the sample. For monodisperse molecules, the output from the concentration-sensitive and molecular-weight-sensitive detectors overlay exactly. As molecular weight polydispersity increases, the output from the molecular-weight-sensitive detectors appears at increasingly lower elution volumes. For example, Figure 3 shows computer simulations of detector tracings for a narrow MWD ($M_w/M_n = 1.05$) and Figure 4 shows tracings for a broad MWD ($M_w/M_n = 2.0$). The shift in the viscometer peak is less than the light-scattering peak because the viscosity is proportional to the molecular weight raised to the power of the exponent in the Mark–Houwink equation $[\eta] = KM_{va}^{9}$, where K and a are empirical constants for a given polymer-solvent system and M_v is the viscosity-

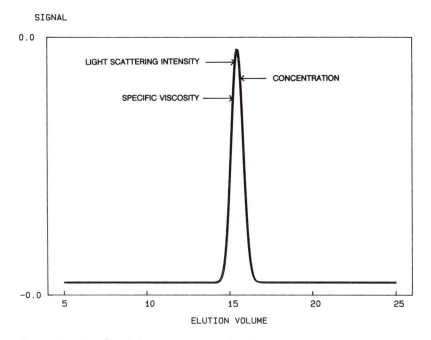

Figure 3.   Simulated detector tracings from light-scattering, viscosity, and refractive index detectors for a narrow MWD polymer. (Reproduced with permission from reference 15. Copyright 1993 Elsevier)

SIGNAL

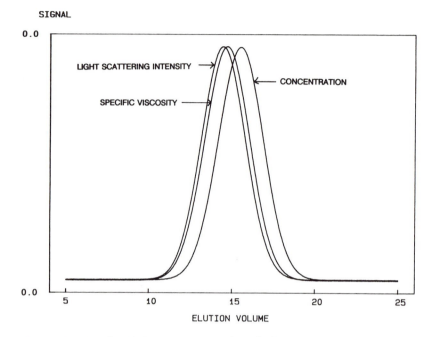

Figure 4.   Simulated detector tracings from light-scattering, viscosity, and refractive index detectors for a broad MWD polymer. (Reproduced with permission from reference 15. Copyright 1993 Elsevier)

average molecular weight. For a random coil conformation, a is less than unity so that the peak from the viscosity detector is shifted by a volume increment related to a *(14)*. The magnitude of the volume shift is comparable with the dead volume between detectors and the volume between consecutive data points. Thus, small errors in determining interdetector volume can lead to significant errors in the measured molecular weight polydispersity.

The interdetector volume can be calculated from the volume of the detector cells and connecting tubing. However, in practice this value has been found to give incorrect results, probably because of viscosity effects and different flow profiles in the different detectors *(10)*. As a result we have to measure an effective volume delay. The volume delay between the viscometer and the refractometer can be determined easily by aligning the peak signals from an injected monodisperse low-molecular-weight solute. Unfortunately such a material will probably not be detected by the light-scattering detector. Narrow MWD polymers can be used, but the small polydispersity may lead to the volume delay being overestimated. One way around this problem is to align the peaks or peak onset of an excluded polymer.

The situation is complicated by the relationship between the inter-detector volume and band broadening. Because of imperfect resolution and band broadening, the polydispersity measured by SEC–light scattering or the intrinsic viscosity distribution measured by SEC–viscometry is narrower than the true polydispersity. If the true polydispersity of a polymer is known, the volume offset can be adjusted so that the measured polydispersity matches the true polydispersity. In effect, a band-broadening correction can be incorporated into the interdetector delay volume to give an "effective" volume offset.

Instrument Parameters.   In conventional SEC, column calibration is required to determine MWDs relative to the calibration standards. Calibration of the concentration detector is usually not required because only relative concentrations are needed. This makes conventional SEC a very precise method as only the elution volume and the relative detector signal are measured. However, the accuracy of the measurement depends on the accuracy of the calibration curve. Although molecular-weight-sensitive detectors avoid some of the accuracy problems associated with column calibration, a number of other calibration procedures become necessary. When a molecular-weight-sensitive detector is added, the absolute concentration of the eluting polymer is needed. To measure this absolute concentration, the concentration detector must be calibrated and the sample property used to measure concentration (refractive index or absorbance) must also be known. In addition, the light-scattering detector, and in general the viscometer also, requires calibration. For light-scattering analysis, the specific refractive index of the polymer in the mobile phase is needed (usually, the second virial coefficient is taken as zero, which does not introduce significant error). For viscometry, the universal calibration curve is required to derive MWDs.

The dependence of molecular weight results on these parameters means that great care must be taken to obtain accurate initial calibration constants. Furthermore, standard reference materials are required to check system operation at regular intervals.

Band Broadening.   The MWD measured using molecular-weight-sensitive detectors is affected by band broadening. However, the effect of band broadening on the measured MWD is different from that of conventional SEC. In SEC the broadening of the peak is interpreted as a broadening of the MWD. However, this broadening is really a loss of resolution caused by sample mixing and imperfect resolution. As a result, the width of the MWD from light scattering and the intrinsic viscosity distribution (IVD) from viscometry is underestimated as compared with a conventional SEC analysis. The weight-average molecular weight and intrinsic viscosity, however, are unaffected if a light-scattering detector or viscometer is used, respectively. If both detectors are used, measurements of molecular conformation and branching based on direct measurement of these quantities are insensitive to band-broadening errors (15, 16). The errors caused by band broadening are generally less than in conventional SEC but may still be significant. However, the errors obtained from universal calibration, whether used to convert the IVD into the MWD or vice versa, are greater than in conventional SEC. Consequently, great care must be taken in establishing adequate band-broadening correction parameters.

Instrument Sensitivity and Baseline Settings. By their nature, instruments measuring different polymer properties will have different sensitivities and measurement ranges (3). These differences need to be considered when evaluating results, especially at the extremes of the distributed properties, where the signal-to-noise ratio is poor (e.g., *see* reference 17). This situation is illustrated in Figure 5, which shows light-scattering and refractometer tracings from a hypothetical MWD. A small amount of high-molecular-weight material is detected by the light-scattering photometer, but the concentration is too low to register on the refractometer. At the low-molecular-weight end of the distribution, the situation is reversed. These effects result in the breadth of the MWD being underestimated.

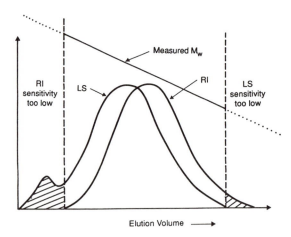

Figure 5. Light-scattering and refractive index detector tracings for a hypothetical MWD containing a small amount of high-molecular-weight material. (Reproduced with permission from reference 3. Copyright 1995 Marcel Dekker)

Finally, multiple signals require multiple baseline settings. Because measured quantities depend on ratios of baseline-corrected signals, it is essential that baselines are set carefully and consistently to maximize the reproducibility of the results. As the measured signals approach the baseline at the peak edges, results based on ratios of these signals become especially error prone, and caution should be exercised when interpreting data from these sections of the chromatogram. This situation can be improved by using Fourier transforms of the measured detector outputs to remove high-frequency noise (*13*).

Conclusions

Multiple detector SEC instrumentation offers great advantages in the accurate characterization of complex polymeric materials, such as copolymers, polymer blends, and branched polymers. However, to generate reproducible results on a routine basis, special care must be taken regarding the added complexity of the instrumentation. In particular, detector configuration should be chosen carefully, interdetector volumes measured precisely, concentration-detector response calibrated, baseline settings and instrument sensitivity parameters should be selected with care, and band-broadening corrections used if needed. Finally, to verify the accuracy of a multiple detector SEC system, the instrument must be evaluated using well-characterized standards.

References

1. Yau, W. W.; Kirkland, J. J.; Bly, D. D. *Modern Size Exclusion Liquid Chromatography*; John Wiley and Sons: New York, 1979.
2. *Steric Exclusion Liquid Chromatography of Polymers*; Jânca, J., Ed.; Marcel Dekker: New York, 1984.
3. Jackson, C.; Barth, H. G. In *Handbook of Size Exclusion Chromatography*; Wu, C. S., Ed.; Marcel Dekker: New York, 1995; p 103.
4. *Detection and Data Analysis in Size Exclusion Chromatography*; Provder, T., Ed.; ACS Symposium Series 352; American Chemical Society: Washington, DC, 1987.
5. Grubisic-Gallot, Z.; Lingelser, J. P.; Gallot, Y. *Polym. Bull. (Berlin)* **1990**, *23*, 389.
6. Lesec, J.; Volet, G. *J. Appl. Polym. Sci., Appl. Polym. Symp.* **1990**, *45*, 177.
7. Lesec, J.; Volet, G. *J. Liq. Chromatogr.* **1990**, *13*, 831.
8. Tinland, B.; Mazet, J.; Rinaudo, M. *Makromol. Chem., Rapid Commun.* **1988**, *9*, 69.
9. Jackson, C.; Barth, H. G.; Yau, W. W. Presented at the Proceedings of the International Gel Permeation Chromatography Symposium, San Francisco, CA, 1991.
10. Haney, M. A.; Jackson, C.; Yau, W. W. Presented at the Proceedings of the International Gel Permeation Chromatography Symposium, San Francisco, CA, 1991.
11. Lecacheux, D.; Lesec, J. *J. Liq. Chromatogr.* **1982**, *12*, 2227.

12. Malihi, F. B.; Kuo, C.; Koehler, M. E.; Provder, T.; Kah, A. F. In *Size Exclusion Chromatography: Methodology and Characterization of Polymers and Related Materials*; Provder, T., Ed.; ACS Symposium Series 245; American Chemical Society: Washington, DC., 1984; p 281.
13. Kuo, C. Y.; Provder, T.; Koehler, M. E.; Kah, A. F. In *Handbook of Size Exclusion Chromatography*; Wu, C. S., Ed.; Marcel Dekker: New York, 1995.
14. Mourey, T. H.; Miller, S. M.; Balke, S. T. *J. Liq. Chromatogr.* 1990, *13*, 435.
15. Jackson, C.; Yau, W. W. *J. Chromatogr.* 1993, *645*, 209.
16. Cheung, P.; Lew, R.; Balke, S. T.; Mourey, T. H. *J. Appl. Polym. Sci.* 1993, *47*, 1701.
17. Pang, S.; Rudin, A. *Polymer* 1992, *33*, 1949.

RECEIVED for review January 6, 1994. ACCEPTED revised manuscript January 17, 1995.

Computer Simulation Study of Multidetector Size-Exclusion Chromatography

Flory–Schulz Molecular Weight Distribution

Christian Jackson and Wallace W. Yau[1]

Central Research and Development, DuPont, Experimental Station, Wilmington, DE 19880–0228

A computer simulation of size-exclusion chromatography–viscometry–light scattering is described. Data for polymers with a Flory–Schulz molecular weight distribution (MWD) are simulated, and the features of the different detector signals are related to the molecular weight and polydispersity of the distribution. The results are compared with previously reported simulated results using a Wesslau MWD.

T HE ACCURACY OF MEASUREMENTS of polymer molecular weight distribution (MWD) by size-exclusion chromatography (SEC) can be improved by the addition of a molecular-weight-sensitive detector, such as an on-line viscometer or light-scattering (LS) detector. These detectors measure solution properties related to molecular weight of the fractionated polymer. Coupling both of these detectors in one SEC instrument potentially offers improved accuracy, precision, and dynamic range for SEC polymer conformation studies (*1–5*). However, the increased complexity of these experiments and the subsequent data handling introduce a number of problems not present in conventional SEC (*6–10*). A computer simulation of multiple detector SEC was developed to study these effects in detail. Two models of the MWD were used: the Wesslau logarithm to the base 10 (log) normal MWD and the Flory–Schulz most probable MWD (*11–14*). The models are described

---

[1] Current address: Chevron Chemical Company, P.O. Box 7400, Orange, TX 77631.

0065–2393/95/0247–0069$12.00/0

and the simulated data are used to illustrate the features of SEC with multiple detectors.

Methodology

Wesslau MWD. The model based on the Wesslau MWD has been described previously (15). The weight fraction distribution of x-mer, where x is the degree of polymerization (DP), measured as a function of the logarithm of the degree of polymerization, is given by

$$w_x = \frac{\ln (10)}{\beta \pi^{1/2} x} e^{(-1/\beta^2 \ln^2 (x/x_0))} \tag{1}$$

where

$$\beta^2 = \ln \left(\frac{x_w}{x_n}\right)^2 \tag{2}$$

where x_n is the number-average DP, x_w is the weight-average DP, and x_0 is the peak value.

Flory–Schulz MWD. The weight fraction of polymer at each degree of polymerization, x, at extent of reaction, p, described by the Flory–Schulz distribution (13, 14) is given by

$$w_x = \frac{a}{x_n \Gamma(a + 1)} \left(\frac{ax}{x_n}\right)^a e^{-ax/x_n} \tag{3}$$

where a is related to the molecular weight polydispersity by

$$\frac{x_w}{x_n} = \frac{a + 1}{a} \tag{4}$$

On a logarithmic molecular weight scale based on SEC separation, equation 4 becomes

$$w_x{}^l = \frac{\ln (10)}{\Gamma(a + 1)} \left(\frac{ax}{x_n}\right)^{a+1} e^{-ax/x_n} \tag{5}$$

which corresponds to the concentration detector signal from the SEC experiment.

The LS detector signal at $0°$ is proportional to the concentration multiplied by the molecular weight at each elution volume and is given by

$$l_{(\theta=0)} = M_0 x w_x{}^l \tag{6}$$

where M_0 is the monomer molecular weight. The viscometer signal is proportional to the intrinsic viscosity multiplied by the concentration. The intrinsic viscosity is given by the Mark–Houwink equation

$$[\eta] = K(x M_0)^\alpha \tag{7}$$

The specific viscosity at each slice is then

$$\eta_{sp} = K(xM_0)^\alpha w_x{}^l \tag{8}$$

In the model there is no interdetector volume difference between the three detector signals.

The SEC has a calibration curve, relating elution volume, V, to molecular weight, M, of the form

$$M(V) = D_1 e^{-D_2 V} \tag{9}$$

where D_1 and D_2 were given values of 15×10^8 and 0.62, respectively.

The two weight-fraction MWDs are illustrated in Figure 1 in which they are plotted as a function of the logarithm of molecular weight. Both distributions shown have a number-average molecular weight of 10,000 g/mol and a polydispersity, M_w/M_n, of 2.

Results and Discussion

Peak Positions. The main results for the Wesslau MWD are summarized for comparison with the results from the Flory–Schulz MWD. The most notable feature is that the tracings from the three detectors (the concentration detector, LS detector, and the viscometer) are all symmetrical Gaussian distributions of equal variance. The only differences between the signals are the relative heights, corresponding to the weight-average molecular weight and intrinsic viscosity of the polymer, and the peak positions. For the LS detector the peak maximum, V_L, is shifted to lower elution volume, corresponding to higher molecular weight, than the concentration signal peak, V_R. The magnitude of the volume shift depends on the sample polydispersity and the slope of the SEC calibration curve, D_2,

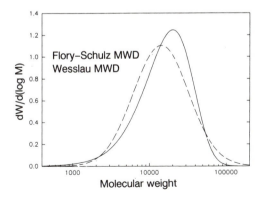

Figure 1. Wesslau and Flory–Schulz differential weight-fraction MWDs on a logarithmic scale, where W is the weight fraction and M is the molecular weight. Both distributions are for M_n = 10,000 g/mol and M_w/M_n = 2.0.

$$V_L = V_R - \frac{\ln\left(x_w/x_n\right)}{D_2} \tag{10}$$

where x_w is weight-average degree of polymerization and x_n is the number-average degree of polymerization. The viscometer peak maximum, V_V, is also shifted to a lower elution volume. For a flexible polymer the shift is less than that for the LS peak. The magnitude of the shift is determined by the value of the Mark–Houwink exponent, a, as well as the polydispersity and the calibration curve slope.

Typical signal tracings for the Flory–Schulz MWD are shown in Figure 2. The tracings have similar shapes but different peak positions. From equation 5 it can be shown that the elution fraction at the maximum in the concentration detector signal has a DP of

$$X_{\text{RI max}} = x_w \tag{11}$$

From equation 6, the LS signal maximum corresponds to the elution fraction with

$$x_{\text{LS max}} = x_w + \frac{x_n}{a}$$

$$= x_z \tag{12}$$

and from equation 8 the viscometer peak is at

$$x_{\text{Visc max}} = (\alpha + a + 1)/ax_n$$

$$= x_w + \frac{\alpha x_n}{a} \tag{13}$$

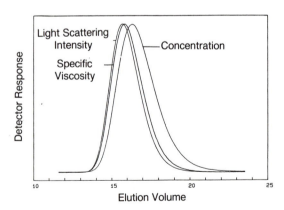

Figure 2. Signal tracings from the three detectors showing excess LS intensity, specific viscosity, and concentration signals, for a sample with a Flory–Schulz MWD, polydispersity of 2, and a Mark–Houwink exponent of 0.725.

In terms of relative peak positions, the LS intensity peak is shifted to an elution volume lower than the concentration detector peak, given by

$$V_L = V_R - \frac{\ln \left((a + 2)/(a + 1)\right)}{D_2} \tag{14}$$

and the viscometer peak is also shifted by a volume given by

$$V_V = V_R - \frac{\ln \left((a + 1 + \alpha)/(a + 1)\right)}{D_2} \tag{15}$$

These volume shifts can be rewritten in terms of the polydispersity, $P = x_w/x_n$, as

$$V_R - V_L = \frac{\ln (P) - \ln (2P - 1)}{D_2} \tag{16}$$

and

$$V_R = V_V = \frac{\ln (P) - \ln (P(\alpha + 1) - \alpha)}{D_2} \tag{17}$$

Rearranging equations 16 and 17 gives the polydispersity in terms of the volume shifts

$$P = \frac{1}{2 - e^{-D_2(V_R - V_L)}} \tag{18}$$

$$P = \frac{\alpha}{1 + \alpha - e^{-D_2(V_R - V_V)}} \tag{19}$$

The relative shifts in peak positions thus depend on the polydispersity of the MWD and the slope of the SEC calibration curve. The shift in the viscometer peak additionally depends on the Mark–Houwink exponent. When the sample is monodisperse, $P = 1$, the signals from all three detectors have the same peak elution volume. If there is any molecular weight polydispersity in the sample, the LS and specific viscosity peaks are shifted to lower elution volumes (higher molecular weight values). The amount of this shift in the LS signal is the measure of the sample polydispersity. In the case of the viscometer, the volume shift depends additionally on the Mark–Houwink exponent. The difference between the viscometer volume shift and the LS volume shift is the measure of the Mark–Houwink exponent. This is the same as for the Wesslau distribution, although the magnitudes of the shifts are different. If the resolution of the chromatograph is increased, the slope of the calibration curve, D_2, will decrease and all the volume differences will increase proportionally. The main information obtained by using a molecular-weight-sensitive detector is the weight-average molecular weight or intrinsic viscosity and the sample polydispersity as shown by the relative position of the detector peaks. As a result, it is critical that the actual physical volume difference that exists between detectors is correctly compensated before data are analyzed.

Peak Shapes. In the case of the Wesslau MWD, the shapes of the peaks from the three detectors are always the same. For the Flory–Schulz distribution, the peak shapes are slightly different and the differences increase with increasing polydispersity. As the polydispersity increases, the LS and viscosity signals become narrower relative to the concentration detector signal and they also become less skewed. Figure 3 shows the peak variance of the viscosity and LS signals relative to the concentration detector peak variance as a function of polydispersity. The concentration detector peak variance increases from 0.25 mL^2 when the polydispersity is 1.1 to 3.65 mL^2 when the polydispersity is 3.3. The LS peak variance increases more slowly. The viscometer variance is in between the two but closer to the LS peak behavior. Figure 4 shows the relative skew of the peaks compared with the refractometer, where the skew is defined as

$$\gamma = \frac{\mu_3}{\mu_2^{3/2}} \tag{20}$$

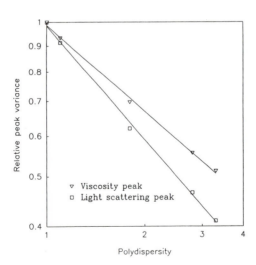

Figure 3.   Variance of the viscosity and LS peaks relative to the variance of the concentration peak as a function of molecular weight polydispersity M_w/M_n.

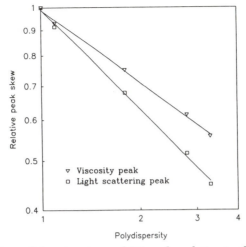

Figure 4.   *Skew of the viscosity and LS peaks relative to the skew of the concentration peak as a function of molecular weight polydispersity* M_w/M_n.

where μ_2 and μ_3 are the second and third moments of the peak, respectively. The behavior is similar to that of the peak variance. The skew of the LS and viscosity peaks increases less with polydispersity than the skew of the concentration signal.

Figures 5 and 6 show the difference in peak shapes in more detail. The LS and concentration tracings are shown as a function of elution volume for distributions with polydispersities of 1.1 (Figure 5) and 2

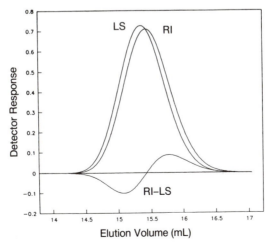

Figure 5.   *Difference between the normalized LS and concentration (RI) signals as a function of elution volume for a MWD with polydispersity 1.1.*

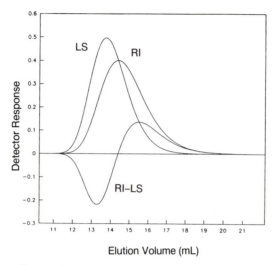

Elution Volume (mL)

Figure 6. Difference between the normalized LS and concentration (RI) signals as a function of elution volume for a MWD with polydispersity 2.

(Figure 6). The signals are normalized so that the two peaks have equal areas, and the difference between these two signals is also plotted. The signals have equal intensity at the peak of the concentration detector response, corresponding to the weight-average molecular weight. There is a maximum in the difference at the number-average molecular weight, and there is a minimum at a molecular weight equal to $4M_n$ corresponding to the $z + 1$ average. The LS peak is higher than the concentration detector peak, and the difference increases with increasing molecular weight. The viscosity peak behaves in a similar way to the LS peak, but with the differences reduced by the exponent of the Mark–Houwink equation.

Conclusions

The computer models described provide a functional simulation of SEC–viscometry–LS analysis of linear polymers. The results for the Flory–Schulz MWD are in qualitative agreement with previous results for the Wesslau MWD. Both models emphasize the importance of determining the correct volume offset between the concentration detector and molecular weight-sensitive detectors. For the Flory–Schulz model, the peak shape, as well as the peak elution volume, can provide information about molecular weight polydispersity. Future work will extend the model to incorporate peak skew and polymer branching.

References

1. Jackson, C.; Barth, H. G. In *Size Exclusion Chromatography Handbook;* Wu, C. S., Ed.; Marcel Dekker: New York, 1995; p 103.

2. Yau, W. W. *Chemtracts* **1990**, *1*, 1.
3. Lesec, J.; Volet, G. *J. Liq. Chromatogr.* **1990**, *13*, 381.
4. Yau, W. W.; Jackson, C.; Barth, H. G. In *Proceedings of the International Gel Permeation Chromatography Symposium*; Waters Associates: Amherst, MA, 1993.
5. Jackson, C.; Barth, H. G.; Yau, W. W. In *Proceedings of the International Gel Permeation Chromatography Symposium*; Waters Associates: Amherst, MA, 1993.
6. Mourey, T. H.; Balke, S. T. In *Chromatography of Polymers: Characterization by SEC and FFF*; Provder, T., Ed.; ACS Symposium Series 521; American Chemical Society: Washington, DC, 1993; p 180.
7. Balke, S. T. In *Modern Methods of Polymer Characterization*; Barth, H. G., Mays, J. W., Eds.; John Wiley and Sons: New York, 1991.
8. Mourey, T. H.; Miller, S. M. *J. Liq. Chromatogr.* **1990**, *13*, 693.
9. Jackson, C.; Barth, H. G. *Trends Polym. Sci.* **1994**, *2*, 203.
10. Yau, W. W.; Kirkland, J. J.; Bly, D. D. *Modern Size Exclusion Liquid Chromatography*; John Wiley and Sons: New York, 1979.
11. Wesslau, H. *Makromol. Chem.* **1956**, *20*, 111.
12. Lansing, W. D.; Kraemer, E. O. *J. Am. Chem. Soc.* **1935**, *57*, 1369.
13. Flory, P. J. *Principles of Polymer Chemistry*; Cornell University: Ithaca, NY, 1953.
14. Rodriguez, F. *Principles of Polymer Systems*; McGraw-Hill: New York, 1982.
15. Jackson, C.; Yau, W. W. *J. Chromatogr.* **1993**, *645*, 209.

RECEIVED for review January 6, 1994. ACCEPTED revised manuscript November 12, 1994.

Eliminating Lag Time Estimation in Multidetector Size-Exclusion Chromatography

Calibrating Each Detector Independently

Kevin G. Suddaby, Ramin A. Sanayei, Kenneth F. O'Driscoll, and Alfred Rudin

Institute for Polymer Research, University of Waterloo, Waterloo, Ontario N2L 3G1, Canada

A method of signal matching in multidetector size-exclusion chromatography (SEC) in which signals are matched according to the hydrodynamic volume of the eluants in each detector is presented. This method is based on an independent calibration curve for each detector and enables signals to be matched through the calibration curves of each detector, thus eliminating the need to estimate additional empirical parameters such as lag times. To correctly calibrate molecular weight-sensitive detectors (such as on-line viscometers or light-scattering detectors), it is necessary to account for the molecular weight distribution (MWD) of the calibration standards. Methods of calibration that account for the MWD of the calibration standards are presented. The utility of applying these methods in multidetector SEC analysis is then demonstrated.

THE USE OF MULTIPLE DETECTORS in size-exclusion chromatography (SEC) greatly increases the power of SEC analysis. However, the problem of matching the signals from the different detectors is introduced. The traditional method of dealing with signal matching in multidetector SEC is to calibrate only one detector. This calibration is then extrapolated to the other detectors by applying an offset (lag time) to represent the physical volume between the various detectors.

The effect of varying this offset between detectors has been examined (1). The molecular weight averages and bulk intrinsic viscosity of samples

0065–2393/95/0247–0079$12.00/0

were found to be insensitive to variation of the offset, but the local (slice) intrinsic viscosities were found to be sensitive to the offset. This was shown by the strong dependence of the Mark–Houwink–Sakaruda K and a values on the offset used. Thus, it is imperative that the signals be correctly matched if the relationship between intrinsic viscosity and molecular weight is to be determined.

The physical interdetector volume can be determined experimentally using marker analysis. However, the error associated with these measurements is quite large; different methods of marker analysis give diverse values for the same interdetector volume (2). Often when the best estimates of the physical interdetector volume are used, unreasonable estimates of the Mark–Houwink–Sakaruda K and a parameters result (1). To obtain reasonable estimates of these parameters, a different offset must be used. In effect the offset between detectors amounts to an adjustable parameter. Clearly, this adjustable parameter is a drawback to using this method of signal matching.

A different approach to signal matching is to calibrate each detector independently. Because each detector is calibrated independently, the signals can be matched through the corresponding calibration curves alone, eliminating the need for offsets.

Experimental Procedures

Experimental molecular weight distributions (MWDs) were determined using an SEC system equipped with a differential refractive index (DRI), a UV visible, and a Viscotek bridge viscometer as detectors. The column eluent passed through the UV detector and then was split evenly between the viscometer and the DRI detectors. All of the detectors were interfaced with a computer for data acquisition at a rate of 1.8207 points per second. One of two column arrangements was used: either a guard column and four 30-cm PLgel 10-μm columns (10^5, 10^3, 500, and 100 Å) or three 30-cm mixed-bed columns (2 PLgel mixed B and 1 Jordi mixed bed). The SEC eluent was tetrahydrofuran (THF), and the system was operated at 30 °C. The columns were calibrated using the standards detailed in Table I. The standards used in this work were purchased from Pressure Chemical Company, Polymer Laboratories Ltd., and Hewlett Packard. The poly(methyl methacrylate) (PMMA) calibration curve was obtained by use of the "universal calibration curve" concept using the $[\eta]$–molecular weight relationship given in equation 2 and appropriate literature values for the parameters (given in text). A multidetector SEC analysis program that implements the concepts presented in this chapter was used throughout this work and is available from the authors.

Independent Detector Calibration

The calibration of SEC requires the use of standard samples. A series of commercially available narrow MWD standards with known peak molecular weights (M_p) is normally used. The M_p value assigned to a cali-

Table I. M_p Values for Polystyrene Standards in THF

Supplied M_p^1	PDI^1	Viscometer M_p^2	LLALLS M_p^2
900	1.16	950	1000
1050	1.21	1120	1180
1770	1.06	1820	1860
2000	1.06	2060	2100
3550	1.05	3640	3700
7600	1.04	7760	7870
17,000	1.04	17,370	17,610
20,400	1.06	21,050	21,440
50,000	1.06	51,710	52,540
97,200	1.06	100,690	102,140
200,000	1.06	207,640	210,170
394,000	1.06	409,850	413,984
600,000	1.10	637,750	646,010
900,000	1.10	958,260	969,150
1,800,000	1.20	1,996,320	2,022,450
2,750,000	1.07	2,892,010	2,909,010
4,250,000	1.06	4,446,360	4,466,180

[1] Nominal M_p and PDI values assigned by the supplier.
[2] Viscometer and LALLS M_p values calculated using a Gaussian MWD and in the case of the viscometer equation 2 (8).

bration standard corresponds to the peak in the response of a differential refractometer (DRI) as this is currently the standard detector in SEC. In practice the elution volume corresponding to the peak in the detector response is determined for each standard and a calibration curve is made. However, because the M_p value supplied with the standard is for a concentration detector (DRI), this method of calibration applies only to the concentration detectors. M_p values for other types of detectors, such as on-line viscometers and light-scattering detectors, are not supplied, making their calibration more complicated.

The response of a concentration detector for the ith slice of the chromatogram is proportional to the (weight) concentration of polymer eluting through the detector, $S_{i,conc} \propto c_i$. The response of other SEC detectors is usually proportional to the product of the concentration and some function of the molecular weight. For example, at the concentrations typically used in SEC, the response of an on-line viscometer is proportional to the product of the concentration and intrinsic viscosity ($[\eta]$) of polymer, $S_{i,visc} \propto c_i[\eta]_i$. Similarly, the response of a low-angle laser light-scattering detector (LALLS) is proportional to the concentration times the molecular weight of polymer, $S_{i,LALLS} \propto c_iM_i$. Because the calibration standards are not truly monodisperse, the molecular weight dependence of these detectors causes the peak in the detector response to be shifted away from the supplied (concentration detector)

M_p. This is illustrated in Figure 1 in which the response calculated for a LALLS detector (by taking the product of concentration and molecular weight of each slice) is superimposed on the measured MWD (DRI, for the standard labeled M_p = 17,000 by the supplier) from which it was generated. The magnitude of this shift depends both on the MWD of the calibration standard and on the molecular weight dependency of the detector response.

The M_p for a concentration detector is the molecular weight corresponding to the peak in the MWD of the polymer. The M_p for an on-line viscometer is the molecular weight corresponding to the peak in the distribution obtained by expressing the product weight fraction times intrinsic viscosity as a function of molecular weight. Similarly, the M_p for a LALLS detector can be obtained by expressing the product weight fraction times molecular weight as a function of molecular weight. These latter two distributions can be generated from the MWD. Because of this variation in M_p, significant errors are introduced if molecular weight-sensitive detectors are calibrated using concentration detector M_p values. These errors can be largely eliminated if the breadth of the MWD of the calibration standards is accounted for in calibrating the molecular weight-sensitive detectors.

Transformation of the MWD

Two methods that determine the correct M_p values of the standards for molecular weight-sensitive detectors are discussed (3). These methods are based on transforming the MWD of the calibration standards into the responses of the molecular weight-sensitive detectors. The first

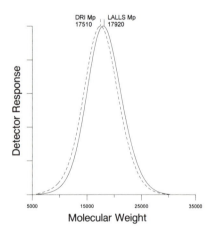

Figure 1. Comparison of the peak molecular weights for a DRI and LALLS detector for the standard labeled M_p *17,000 by the supplier.*

method is an experimental one that uses the MWD of the standard as determined from the SEC concentration chromatogram. The second method is a theoretical treatment of the data that approximates the MWD of the standards with a known distribution. In both methods the M_p values of the molecular weight-sensitive detectors are determined from the peak of the transformed distribution. These M_p values are then used to obtain the correct calibration for each detector.

The transformation of the experimentally determined MWD is relatively straightforward. The concentration of each species in the standard sample, c_i, is multiplied by its intrinsic viscosity, $[\eta]_i$, to simulate the viscometer signal. Similarly, the LALLS signal is simulated by multiplying the concentration of each species in the sample, c_i, by its molecular weight, M_i. In generating the viscometer response, an established intrinsic viscosity–molecular weight relationship for the SEC solvent and calibration polymer should be used. The molecular weights used are obtained from the concentration detector calibration. This method was used to calibrate the viscometer for the samples discussed later in this chapter.

The alternative to transforming the experimentally determined MWD is to transform an assumed MWD. Although any form can be used for the distribution, the form used should be such that it can be expressed in terms of parameters supplied with the calibration standards. Both log normal and Guassian distributions have been used to approximate the MWD of unimodal polymers (4). Both distributions were considered here, and statistically the Gaussian distribution was found to better approximate the MWDs of the narrow MWD standards used in this work. As a result the following Gaussian distribution was used.

$$w(M_i) = \frac{\exp\left(\dfrac{-(M_i - M_p)^2}{2\left(\dfrac{M_p^{\,2}}{\mathrm{PDI}} - \dfrac{M_p^{\,2}}{\mathrm{PDI}^2}\right)}\right)}{\sqrt{2\pi\left(\dfrac{M_p^{\,2}}{\mathrm{PDI}} - \dfrac{M_p^{\,2}}{\mathrm{PDI}^2}\right)}} \tag{1}$$

This distribution expresses the weight fraction of polymer of a given molecular weight, $w_i(M_i)$, as a function of two parameters, M_p and polydispersity (PDI) (PDI = M_w/M_n) both of which are supplied with commercial standards. Thus, the MWD of the calibration standards can be approximated from data supplied with the standards. The transformations discussed previously can be applied to the assumed MWD and estimates of the M_p values for the molecular weight-sensitive detectors can be obtained.

Each of the two methods proposed for determining M_p for molecular weight-sensitive detectors has advantages and disadvantages. Using the experimentally determined MWDs of the calibration standards has the advantage of making no assumptions about the nature of the MWD of the standards. Also, because the MWD is determined using information from all the calibration standards (because the concentration detector calibration curve is required to obtain the MWD), it is less susceptible to errors in the individual values of M_p and PDI assigned by the supplier. A slight disadvantage occurs because the M_p values of the molecular weight-sensitive detectors are determined using the concentration detector; therefore, the calibration curves of the different detectors lose some of their independence. Closely related to this is the fact that this method yields system-dependent M_p values. Also, in systems where band broadening is significant, the MWDs of the standards will be subject to this effect. However, because the calculated M_p values are heavily weighted by the central parts of the distribution, the values are expected to be largely insensitive to peak broadening.

The most significant advantage of assuming a form for the MWDs of the standards is that it results in M_p values that are independent of the system calibration. However, this method is subject to errors arising from the assumption of a form for the MWD. Also, because the data from each standard alone are used in this method, it is subject to errors in the values of the parameters supplied with the calibration standards.

Significance of Peak Molecular Weight Corrections for Molecular Weight-Sensitive Detectors

Table I quantifies the effect illustrated in Figure 1 for a series of polystyrene standards in THF. In this table the M_p values for the molecular weight-sensitive detectors were obtained by assuming the calibration standards have a Gaussian MWD. It is apparent that despite the low polydispersities of the calibration standards (typically PDI ≈ 1.06), the M_p values of the molecular weight-sensitive detectors are significantly different from the M_p values assigned by the supplier. For example, the standard assigned a value of 200,000 for $M_{p,conc}$ and a PDI of 1.06 by the supplier is predicted to have values of 207,640 and 210,170 for $M_{p,visc}$ and $M_{p,LALLS}$, respectively. The effect of polydispersity on $M_{p,LALLS}$ is larger than the effect on $M_{p,visc}$ because of the difference in the molecular weight sensitivity of the two detectors. The LALLS signal scales as M, whereas the viscometer signal scales (through $[\eta]$) approximately as $M^{0.7}$ in good solvents.

Both the experimental and theoretical methods of determining the correct M_p values for the molecular weight-sensitive detectors give similar M_p values for calibration standards. This fact is apparent from Figure

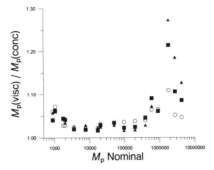

Figure 2. *Ratio of the viscometer peak molecular weight to the concentration detector peak molecular weight versus* M_p *value supplied with the standards. Open circles were determined assuming a Gaussian MWD for the standards, squares were determined from the DRI MWD, and triangles were determined from the UV MWD.*

2, which shows the ratio of $M_{p,visc}$ to $M_{p,conc}$ (the viscometer M_p shift) versus the nominal M_p for the series of standards given in Table I. The values determined from both the experimental MWDs (filled symbols) and the Gaussian approximation (circles) are shown. The points corresponding to the experimentally determined MWDs include data from both a DRI detector (filled squares) and a UV detector (filled triangles). The SEC system used is equipped with both of these detectors, which act as concentration detectors for polystyrene. Figure 3 is a similar plot that shows the LALLS M_p shift for the same series of standards.

Figures 2 and 3 show that the M_p corrections determined by the experimental and theoretical methods are comparable. The corrections

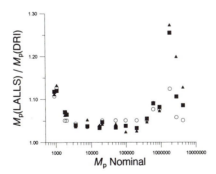

Figure 3. *Ratio of the LALLS peak molecular weight to the concentration detector peak molecular weight versus* M_p *value supplied with the standards. Open circles were determined assuming a Gaussian MWD for the standards, squares were determined from the DRI MWD, and triangles were determined from the UV MWD.*

vary from a few percent in the middle of the molecular weight range to over 10% at the molecular weight extremes. Because the uncertainties in typical SEC analyses are of the order of 5–10%, these corrections are significant. The trends in the magnitude of the corrections result from the higher polydispersity of the standards at the molecular weight extremes.

Figure 4 demonstrates the importance of using the correct M_p values when calibrating molecular weight-sensitive detectors. This figure shows the ratio of the M_n determined from the viscometer alone (5, 6) to the M_n determined from the DRI detector for a series of polystyrene standards. The lower dataset was obtained by calibrating the viscometer with the uncorrected M_p values supplied with the standards. It is apparent in this case the viscometer consistently underestimates M_n relative to the DRI. The upper dataset was obtained using the correct M_p values for calibrating the viscometer. This dataset is scattered about 1, indicating that there is good agreement between the M_n from the viscometer and M_n from the DRI when the nature of the viscometer response is taken into account during its calibration.

Matching Detector Signals

The signals from independently calibrated detectors can be matched through their respective calibration curves. The signals from the two detectors are matched at a given hydrodynamic volume. This is illustrated schematically in Figure 5 in which the on-line viscometer and DRI signals are combined to obtain an $[\eta]$-MW trace for a sample. Each slice of the concentration (DRI) chromatogram can be represented by an elution volume V_1 and a concentration c_1. Using the DRI

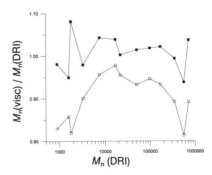

Figure 4.   Comparison of M_n *from the viscometer to* M_n *from the DRI detector. Open squares were obtained using the* M_p *values supplied with the standards (concentration detector* M_p*) in calibration. Filled squares are obtained when viscometer* M_p *values determined by the experimental method detailed in the text are used in calibration.*

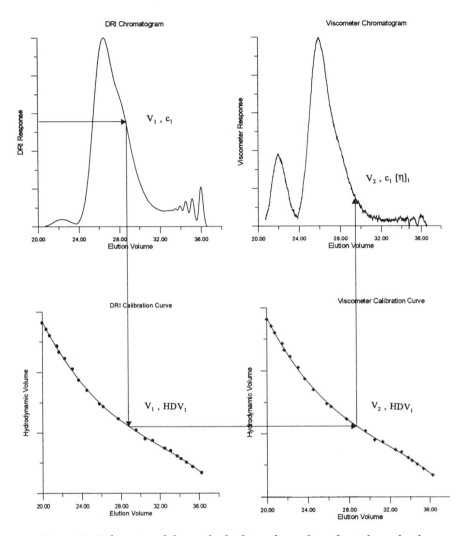

Figure 5. Schematic of the method of signal matching for independently calibrated detectors (see text).

calibration curve, this elution volume can be transformed into the corresponding hydrodynamic volume HDV_1. This hydrodynamic volume can then be used with the viscometer calibration curve to determine the corresponding viscometer elution volume V_2. The viscometer elution volume is then used with the viscometer chromatogram to determine the viscometer signal that corresponds to the slice of the DRI chromatogram $c_1[\eta]_1$. By dividing the viscometer signal by the corresponding DRI signal, the intrinsic viscosity of the slice can be determined. This procedure is then repeated across the chromatogram.

In the preceding procedure, there is in effect an "offset" between the detector signals (the difference between the elution volumes V_1 and V_2 at a given hydrodynamic volume). However, there is no need to estimate this offset because this information is contained in the independent calibration curves. Because the signals are matched through their independent calibration curves, the offset is not necessarily static and, depending on the calibration curves, the offset may vary across the chromatogram.

It is possible to calculate the offsets between detectors from the independent calibration curves. This was done and the results are shown in Figure 6. The calculated offset between the DRI and UV detectors is constant within experimental uncertainty. In contrast, the calculated offset between the viscometer and DRI detectors shows deviations at its extremes. The cause of these deviations is unclear, but they may be a manifestation of flow effects such as those observed in systems with single capillary viscometers (7).

Application

Figure 7 is a plot of the [η]-MW relationship for a multimodal PMMA (M_n 3080, M_w 19,120) superimposed onto the MWD of the sample. The smooth bold curve in this figure represents the literature intrinsic viscosity molecular weight relationship

$$[\eta]_i = K_\theta M_i^{1/2} + K'M_i \tag{2}$$

given by equation 2 with values of 7.3×10^{-2} mL/g for K_θ and 1.12×10^{-4} mL/g for K' (8). The noisier curve displays the slice intrinsic

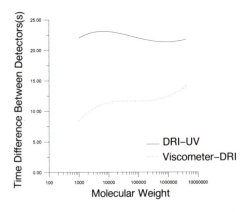

Figure 6.   Differences in peak elution times as a function of molecular weight calculated from the polystyrene calibration curves for each detector.

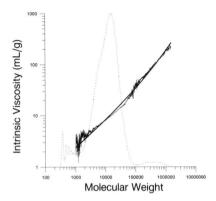

Figure 7.   Intrinsic viscosity versus molecular weight plot for a multimodal PMMA.

viscosities that were determined by the method described in the previous section. The agreement between the predicted and measured intrinsic viscosities is satisfying because the sample is multimodal and covers three orders of magnitude in molecular weight with widely varying concentrations. It is interesting to note that deviations between the slice intrinsic viscosity and literature intrinsic viscosity occur where the weight fraction in the eluent undergoes a marked change. This is perhaps due to mixing effects in the detector cells and needs further investigation.

Equation 2 is used for the $[\eta]$-MW relationship because this equation has been shown to be valid over a much wider molecular weight range than a single set of Mark–Houwink–Sakaruda parameters (8). The molecular weight range covered by this PMMA sample is too wide to be described by the Mark–Houwink–Sakaruda equation.

Figure 8 is a plot of the $[\eta]$-MW relationship for a very broad MWD polystyrene (M_n 17,290, M_w 77,620). The smooth bold curve represents the relationship given by equation 2 with K_θ 8.5 × 10^{-2} mL/g and K' 1.74 × 10^{-4} mL/g, the literature values for polystyrene (8). The noisier curve displays the experimentally determined slice intrinsic viscosities. Again, there is good agreement between the experimental and predicted results. The dashed curve represents the Mark–Houwink–Sakaruda equation for polystyrene in THF (K 1.47 × 10^{-2} mL/g, a 0.702) (1), which is clearly not valid over the whole of the wide molecular weight range covered by this sample.

Conclusions

Valid independent calibration curves can be established for each detector in a multidetector SEC system. In doing so the nature of the detector response and the MWDs of the calibration standards must be taken into

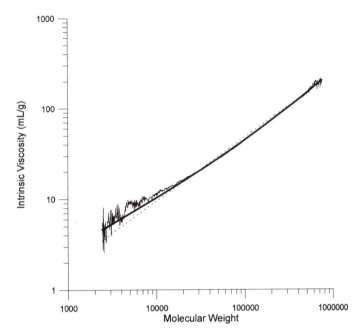

Figure 8. Intrinsic viscosity versus molecular weight plot for a broad MWD polystyrene.

account during the calibration. This method enables the detector signals to be matched based on the hydrodynamic volume of the eluants passing through each detector. Using this method gives reliable results and also eliminates the need to use an adjustable parameter such as a time lag between detectors.

Acknowledgment

Support for this research by the Natural Science and Engineering Research Council of Canada is gratefully acknowledged.

References

1. Kuo, C.; Provder, T.; Koehler, M. E. In *Chromatography of Polymers: Characterization by SEC and FFF*; Provder T., Ed.; ACS Symposium Series 521; American Chemical Society: Washington, DC, 1993; pp 231–242.
2. Balke, S. T.; Cheung, P.; Jeng, L.; Lew, R.; Mourey, T. H. In *Polymer Analysis and Characterization. III*; Barth, H. G.; Janca, J. Eds.; Applied Polymer Symposia 48; John Wiley & Sons: New York, 1991; pp 259–291.
3. Suddaby, K. G.; Sanayei, R. A.; O'Driscoll K. F.; Rudin, A. *Makromol. Chem.* 1993, *194*, 1965–1974.
4. Billingham, N. C. In *Comprehensive Polymer Science*; Eastmond, G. C.; Ledwith, A.; Russo, S.; Sigwalt, P., Eds.; Pergamon: Oxford, England, 1989; p 43.

5. Sanayei, R. A.; Suddaby, K. G.; Rudin, A. *Makromol. Chem.* **1993**, *194*, 1953–1963.
6. Goldwasser, J. M. In *Chromatography of Polymers*; Provder T., Ed.; ACS Symposium Series 521; American Chemical Society: Washington, DC, 1993; pp 243–251.
7. Lesec, J.; Millequant, M.; Havard, T. In *Chromatography of Polymers: Characterization by SEC and FFF*; Provder T., Ed.; ACS Symposium Series 521; American Chemical Society: Washington, DC, 1993; pp 220–230.
8. Sanayei, R. A.; O'Driscoll, K. F.; Rudin, A. In *Chromatography of Polymers: Characterization by SEC and FFF*; Provder T., Ed.; ACS Symposium Series 521; American Chemical Society: Washington, DC, 1993; pp 103–112.

RECEIVED for review January 6, 1994. ACCEPTED revised manuscript July 19, 1994.

8

Determination of Molecular Weight and Size of Ultrahigh Molecular Weight Polymers Using Thermal Field-Flow Fractionation and Light Scattering

Seungho Lee and Oh-Seung Kwon

3M Company, 3M Center, St. Paul, MN 55144

Thermal field-flow fractionation was used to characterize ultrahigh molecular weight poly(methyl methacrylate) (PMMA) with a light-scattering detector. The influence of the differential refractive index increment (dn/dc) and the second virial coefficient (A₂) on the measured molecular weight (MW) and the molecular size was investigated using a broad polystyrene standard having the nominal MW of 250,000. No significant change was observed in MW, and size with A₂–A₂ could be assumed to be zero. Debye plot showed a good linearity for the entire range (0–180°) of the scattering angle. For ultrahigh MW PMMA, Debye plot was not linear, and the multiangle measurement was necessary for the extrapolation of data. Both MW and size increased with A₂, and thus A₂ could not be assumed to be zero for ultrahigh MW polymers.

POLY(METHYL METHACRYLATE) (PMMA) has remained the most widely used material for the optic portion of the intraocular lens since it was first implanted into human eyes in the late 1940s. It is dimensionally and chemically stable and more transparent than most other types of optical glasses. The original type very high molecular weight (MW) PMMA. In this form, PMMA is amenable to lathe-cutting, compression-casting, and cast-molding fabrication techniques. It can also be tumble-polished. Because of PMMAs excellent balance of properties, very little has been done until recently to develop other optic materials. Generally, acrylic polymers are brittle. Modified acrylics having properties unattainable by the basic unmodified compositions are now offered (*1–3*).

0065–2393/95/0247–0093$12.00/0
© 1995 American Chemical Society

MW determination of these high MW polymers using size-exclusion chromatography (SEC) is sometimes difficult because portions of these materials are beyond the linear calibration range of most SEC columns. Thermal field-flow fractionation (ThFFF) has been used for the characterization of a wide range of organic-soluble polymers in dilute solutions (4). In ThFFF, elution volume of a sample is a function of MW, and thus the molecular weight distribution (MWD) of the sample can be determined from its elution profile. ThFFF offers higher resolution than SEC in its normal range of operation (5). ThFFF is particularly useful for characterizing very high MW polymers that are difficult to analyze using SEC (6–8) and microgel-containing polymers (8).

ThFFF elution volume (or time) is a function of D_T/D, where D_T is the thermal diffusion coefficient and D is the mass diffusion coefficient (9, 10). The mass diffusion coefficient D of a polymer molecule in a fluid with viscosity η_o is given by (11)

$$D = \frac{RT}{6\pi\eta_o N_A} \left(\frac{10\pi N_A}{3[\eta]M}\right)^{1/3} \tag{1}$$

where R is the universal gas constant, T temperature, N_A the Avogadro's number, M viscosity-average MW, and $[\eta]$ intrinsic viscosity. If the value of D_T is available, the MW of a polymer can be determined directly from its ThFFF elution volume using equation 1. Values of D_T are not readily available and no theory exists to describe D_T with known physicochemical parameters. A calibration is usually required to determine MWD of polymers using ThFFF. It is noted that ThFFF has been used to study thermal diffusion phenomenon and to determine D_T (12, 13).

MW-sensitive detectors (e.g., differential viscometer, light scattering detector) have been used to eliminate the need for calibration in ThFFF. With a differential viscometer, the intrinsic viscosity distribution (IVD) of a polymer is measured. The IVD is then converted to MWD using Mark–Houwink (M–H) constants (14). The use of accurate M–H constant is essential in this method. Low-angle laser light scattering (LALLS) has also been used for ThFFF (15). Unlike viscometry, the light-scattering method measures the absolute MW of polymers directly. In multiangle laser light scattering (MALLS), the scattered light intensity is measured over a broad range of the scattering angles. Besides the MW, the molecular size can be measured from the angular dependence of the scattered light intensity. Although MALLS has been used in SEC for various applications (16–18), it has not yet been used with ThFFF. In this study, MALLS was combined with ThFFF to investigate the applicability of ThFFF for the characterization of ultrahigh MW polymers.

Theory

Theories on ThFFF and light scattering have been discussed in numerous publications. Equations that are needed for the discussion of results are briefly reviewed here. In ThFFF, the retention ratio, R, is given by (6)

$$R = \frac{V^o}{V_r} = 6\lambda \tag{2}$$

for well-retained solutes. The full expression for R is somewhat complicated (19) and is not discussed here. V^o is the channel volume and V_r is the elution volume. The retention parameter λ is related to D_T/D by

$$\frac{1}{\lambda} = \left(\frac{D_T}{D}\right)\Delta T \tag{3}$$

where ΔT is the temperature drop across the channel. Combining equations 2 and 3, D_T/D can be calculated from the measured elution volume V_r. The MW is then determined using a calibration curve ($\log (D/D_T)$ vs. $\log M$) constructed with a series of narrow standards. As MW increases, D decreases (eq 1), and λ decreases (eq 3). Thus in ThFFF, low MW species elute earlier than high MW species.

When light passes through an inhomogeneous medium such as a polymer solution, it is scattered in all directions. The light scattering at an angle θ by the solute is measured by the excess Rayleigh ratio R_θ which is defined by

$$R_\theta = f_{geom}\frac{(I_{\theta,solution} - I_{\theta,solvent})}{I_o} \tag{4}$$

where $I_{\theta,solution}$ and $I_{\theta,solvent}$ are the intensities of the scattered light by the solution and the solvent, respectively, and I_o is the intensity of the incident light. The geometric factor $f_{geom} = r^2/V$, where r is the distance between the scattering source and the detector and V is the scattering volume. For a dilute polymer solution, the excess Rayleigh ratio R is related to the weight-average molecular weight (M_w) and the second virial coefficient (A_2) of the polymer by (20)

$$\frac{R_\theta}{K^*c} = M_w P_\theta(1 - 2A_2 c M_w P_\theta) \tag{5}$$

where K^* is a constant defined by (18)

$$K^* = \frac{4\pi^2(dn/dc)^2 n_o^2}{N_A \lambda_o^4} \tag{6}$$

for vertically polarized light, dn/dc is the differential refractive index (RI) increment, n_o is the RI of the solvent at the incident wavelength λ_o. The solute concentration c (g/mL) is calculated by

$$c = \frac{\Delta n}{(dn/dc)} \qquad (7)$$

where Δn is the difference in RI between the solution and the pure solvent. $P\theta$ is the scattering factor (or form factor) and is expressed as a power series in $\sin^2 (\theta/2)$ as

$$P_\theta = 1 - \alpha_1 \sin^2 \left(\frac{\theta}{2}\right) + \alpha_2 \sin^4 \left(\frac{\theta}{2}\right) - \alpha_3 \sin^6 \left(\frac{\theta}{2}\right) + \cdots \qquad (8)$$

which can be simplified to

$$P_\theta = 1 - \alpha_1 \sin^2 \left(\frac{\theta}{2}\right) \qquad (9)$$

for low scattering angles. The coefficient $\alpha_1 = (4\pi n_o/\lambda_o)^2 \langle r_g^2 \rangle_z/3$, where $\langle r_g^2 \rangle_z$ is the z-average mean square radius of the polymer. The root mean square radius (or RMS radius) $\sqrt{\langle r_g^2 \rangle_z}$ is sometimes called "radius of gyration".

For each slice of the fractogram, the intensity of the scattered light is measured at a set of discrete scattering angles, and a Debye plot [R_θ/K^*c vs. $\sin^2 (\theta/2)$] is constructed. When $\theta = 0$, $P\theta = 1$, and equation 5 becomes

$$\left(\frac{R_\theta}{K^*c}\right)_{\theta=0} = M_w (1 - 2A_2 c M_w) \qquad (10)$$

where $(R_\theta/K^*c)_{\theta=0}$ is the y-intercept of the Debye plot. By solving equation 10, M_w of the slice is obtained from

$$M_w = \frac{2 \, (\text{intercept})}{1 + \sqrt{1 - 8A_2 c \, (\text{intercept})}} \qquad (11)$$

It is noted that the product, $2A_2 \, c \, M_w$ is much smaller than 1 for most ThFFF (or SEC) experiments, and equation 10 is further simplified to

$$\left(\frac{R_\theta}{K^*c}\right)_{\theta=0} = M_w \qquad (12)$$

Thus, M_w is directly obtained from the y-intercept of the Debye plot. The mean square radius is obtained from the slope of the Debye plot by

$$\langle r_g^2 \rangle_z = \frac{-3\lambda_o^2 m_o}{16\pi^2 n_o^2 M_w (1 - 4A_2 c M_w)} \tag{13}$$

where m_o is the slope of the Debye plot at zero scattering angle, $m_o = d(R_\theta/K^\circ c)/d(\sin^2 (\theta/2))_{\theta=0}$.

Experimental Details

ThFFF. ThFFF was carried out with a Polymer Fractionator model T100 (FFFractionation, Inc., Salt Lake City, UT) equipped with a Waters model 590 pump and a 20-mL loop Rheodyne injector (Rheodyne Inc., Cotati, CA). Detectors were a MALLS (Wyatt Technology model DAWN-F, Santa Barbara, CA) and a refractive index (RI; Hewlett Packard model 1037A, Palo Alto, CA) connected in series with the RI following the MALLS. The light source of the MALLS is a He–Ne laser (632.8 nm). The ThFFF channel is 0.0127 cm thick, 1.9 cm in breadth, and 45.6 cm long. Light-scattering data were collected and processed using the ASTRA software provided by Wyatt Technology. For conventional ThFFF experiments (without a light-scattering detector), a calibration curve (log (D/D_T) vs. log M) was constructed using a series of narrow polystyrene (PS) standards having MWs of up to 9.35 million Da. The calibration curve showed an excellent linearity for the entire MW range. The ThFFF–RI (refractive index) data were collected and processed using the software provided by FFFractionation, Inc.

dn/dc Measurement. A laser differential refractometer (LDC/Milton Roy model KMX-16, Riviera Beach, FL) was used for dn/dc measurements.

SEC. SEC was carried out at room temperature with an HP 1090 Chromatograph (Hewlett Packard, Palo Alto, CA) equipped with an RI detector (Hewlett Packard model 1037A). Columns were Permagel 500-, 10^3–10^6-, and 100-Å columns (Column Resolution, Inc., San Jose, CA) connected in series. All SEC experiments were run in tetrahydrofuran (THF) at 1.0 mL/min. Samples were dissolved in THF and filtered through a 0.2-mm poly(tetrafluoroethylene) disposable filter (nonsterile, 25 mm disc). The injection volume was 100 mL. The column set was calibrated using a series of narrow PS standards having MW ranging from 2000 to 7.7×10^6. The calibration curve (log MW vs. retention time) was obtained by fitting the data with a third-order linear regression, and the curve started deviating from linearity at the MW of ~5 million due to the column exclusion.

Materials. Narrow PS standards were obtained from Pressure Chemical Company (Pittsburgh, PA). A broad PS standard (MW = 250,000) was obtained from American Polymer Standards Corp. (Mentor, OH). The PMMA materials were Perspex CQ UV obtained from Imperial Chemical Industries (Wilmington, DE) and UV 52E obtained from Pharmacia Ophthalmics Inc.(Monrovia, CA). High-performance liquid chromatography-grade THF from JT Baker Inc. (Phillipsburg, NJ) was used as a carrier for all ThFFF and SEC experiments. The polymer solutions had concentrations of ~0.2% (wt/vol).

Results and Discussion

The determination of polymer MW and RMS radius using light scattering measurement requires the knowledge of dn/dc and A_2 values (*see* eqs 6, 7, 11, and 13). A broad PS standard having the nominal MWs of M_n = 100,000, M_w = 250,000, M_z = 430,000 was used to review the basics of polymer characterization using ThFFF–MALLS–RI. Figure 1 shows ThFFF elution curves of the PS standard obtained from light scattering (at 90°) and RI detector. ThFFF conditions were ΔT = 50 °C, flow rate = 0.3 mL/min with the stop-flow time of 1 min.

The values of dn/dc and A_2 are available in literature for many polymer–solvent systems. For the polymer–solvent systems whose dn/dc and A_2 values are not available, separate measurements are required. The dn/dc can be measured by differential refractometry and A_2 by the static mode of light scattering. A range of dn/dc and A_2 values are reported for the PS–THF system: dn/dc = 0.186 − 0.193 cm³/g and A_2 = 8.32 × 10^{-4} − 2.11 × 10^{-4} mL mol/g² at the wavelength of 633 nm (21). For the PS standard used in study, the dn/dc value of 0.190 was measured using a laser differential refractometer. The A_2 value was not measured separately.

Table I shows the MWs and the RMS radii of the PS standard determined with different dn/dc and A_2 values. Equations 11 and 13 indicate that both the MW and the RMS radius increase as A_2 increases. No significant changes were observed in MWs and sizes when A_2 was varied at the constant dn/dc value of 0.2. It is noted that A_2 could be assumed to be zero. As mentioned earlier, the product A_2cM is usually

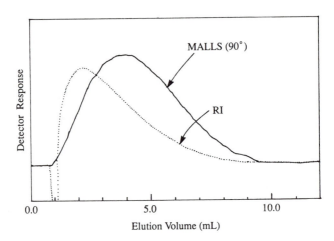

Figure 1. ThFFF elution curves of a broad PS standard having nominal MWs of M_n = 100,000, M_w = 250,000, and M_z = 400,000. ThFFF conditions *are* ΔT = 50 °C *and flow rate* = 0.3 mL/min.

Table I.   Molecular Weights and RMS Radii of 250,000 MW PS Standard
Determined by ThFFF–MALLS–RI Using Different dn/dc and A_2 Values

dn/dc	A_2	M_n	M_w	M_z	$r_{g,n}$ (nm)	$r_{g,w}$ (nm)	$r_{g,w}$ (nm)
0.200	0	1.48×10^5	2.58×10^5	3.99×10^5	15.2	20.0	25.0
0.200	1×10^{-4}	1.48×10^5	2.58×10^5	3.99×10^5	15.2	20.0	25.0
0.200	2×10^{-4}	1.49×10^5	2.58×10^5	4.00×10^5	15.2	20.0	25.0
0.200	4×10^{-4}	1.49×10^5	2.58×10^5	4.00×10^5	15.2	20.0	25.0
0.200	8×10^{-4}	1.50×10^5	2.59×10^5	4.01×10^5	15.2	20.0	25.0
0.200	1×10^{-4}	1.48×10^5	2.58×10^5	3.99×10^5	15.2	20.0	25.0
0.190	1×10^{-4}	1.56×10^5	2.71×10^5	4.20×10^5	15.2	20.0	25.0
0.180	1×10^{-4}	1.65×10^5	2.87×10^5	4.44×10^5	15.2	20.0	25.0

much smaller than 1, and the A_2 terms in equations 11 and 13 are negligible. The term $A_2 cM$ becomes increasingly important as the MW or the concentration of the sample increases. The effect of A_2 on the MW and size for ultrahigh MW PMMA materials is discussed later.

The MW is inversely proportional to the product K^*c (*see* eq 12) and the product K^*c is proportional to dn/dc (*see* eqs 6 and 7). Thus for a given R_θ, the calculated MW is inversely proportional to the dn/dc value used for the calculation. As dn/dc decreases from 0.2 to 0.18 at the fixed A_2 value of 1×10^{-4}, the MWs increase proportionally as expected. Because the size is determined from the slope of the Debye plot, it is independent of dn/dc.

Figure 2 shows the Debye plot for the slice at the elution volume of 3.7 mL. The values of dn/dc and A_2 were taken as 0.19 and zero,

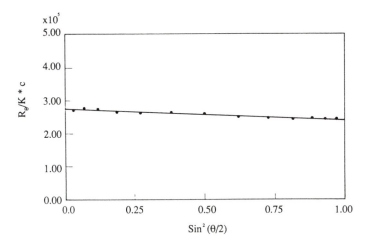

Figure 2.   Debye plot for the PS standard shown in Figure 1 at the elution volume of 3.7 mL.

respectively. The data were extrapolated using the first-order linear regression, and MW of 2.74×10^5 and RMS radius of 20.9 nm were obtained for the slice (*see* eqs 11 and 13). The first-order least square line is shown as a solid line. According to equation 9, the Debye plot is linear at low scattering angles. Figure 2 shows a good linearity for the entire range (0–180°) of the scattering angles.

The M_w and RMS radius determined for the whole distribution of the PS standard are plotted against the elution volume in Figures 3 and 4, respectively. Figure 3 shows a good linearity between log MW and log V_r as expected from ThFFF theory (*see* eqs 1–3). At the beginning and the end of the elution curve, the detector signal becomes too weak to measure the MW accurately. The noise shown at the high end of the elution volume is due to the weak RI response (relative to the light-scattering signal) as it approaches the baseline (*see* Figure 1). Figure 4 also shows the expected increase in size with increasing elution volume. The plot becomes noisy at the elution volume below about 3 mL. As the molecular size becomes much smaller than the wavelength of the light source, the angular dependence of the light scattering disappears (isotropic scattering) and an accurate determination of molecular size becomes difficult.

Figure 5 shows the plot of RMS radius versus MW on a log–log scale. The data for the elution volume lower than 3 mL were dropped

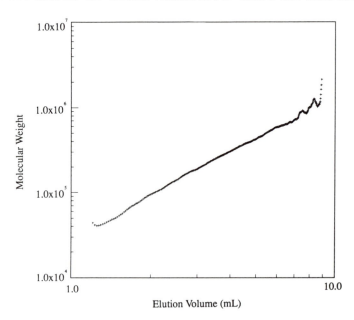

Figure 3. MW versus elution volume for the PS standard shown in Figure 1.

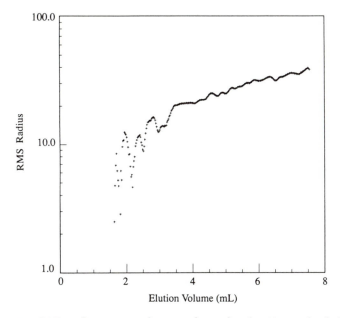

Figure 4. RMS radius versus elution volume for the PS standard shown in Figure 1.

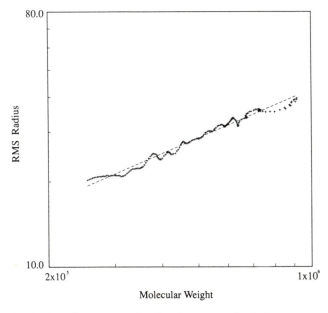

Figure 5. RMS radius versus MW for the PS standard shown in Figure 1.

because of excessive noise in the RMS radius as shown in Figure 4. The slope of the plot depends on the molecular density (*17, 18*). As the density of the molecule increases due to branching, cross-linking, or the existence of the microgels, and so on, the slope of the plot decreases, and thus the RMS radius versus MW plot can be used for polymer conformation studies (*17*). The first-order least-square fit of the data is shown as a dotted line. The slope is 0.57, which agrees well with the result reported elsewhere for PS–THF system (*18*).

The same ThFFF–MALLS–RI system was used to characterize two ultrahigh MW PMMA materials (Perspex and UV 52E). A power programming (*22*) was used for ThFFF operations with the programming parameters of initial ΔT = 40 °C, predecay time t_1 = 5 min, t_a = −10 min, and the hold ΔT = 10 °C. The flow rate was fixed at 0.5 mL/min. Figure 6 shows the traces from light-scattering (90°) and RI detector for the PMMA materials. Figure 7 shows the Debye plot of the Perspex at the elution volume of 11 mL. Unlike the plot for the PS standard shown in Figure 2, the plot is not linear: a fourth-order regression was required to fit the data. As MW increases, the A_2 term (A_2cM) and thus the higher order terms in equation 8 become increasingly important,

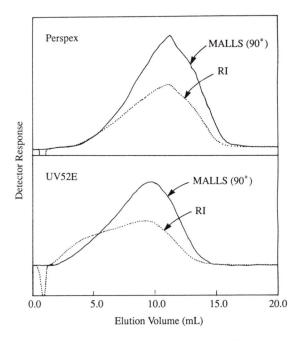

Figure 6.   Elution curves for power programmed ThFFF runs of PMMA materials with parameters t_1 = 5 min, t_a = −10 min, initial ΔT = 40 °C, and flow rate = 0.5 mL/min.

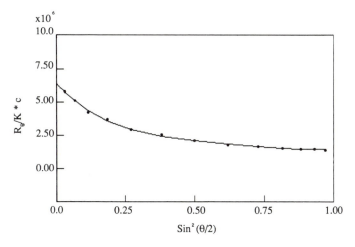

Figure 7.   Debye plot for PERSPEX at the elution volume of 11 mL.

and the Debye plot starts deviating from the linearity. Multiangle measurement is thus necessary for the accurate determination of MW and RMS radius of ultrahigh MW polymers.

A range of dn/dc and A_2 values are also reported for the PMMA–THF system: dn/dc = 0.084–0.088 cm^3/g and A_2 = 5.3 × 10^{-4} to 1.1 × 10^{-4} mL mol/g^2 at the wavelength of 633 nm (21). Table II shows the MWs and the RMS radii calculated for the Perspex using different dn/dc and A_2 values. Unlike for the PS standard discussed earlier, both MW and RMS radius increase with A_2. The A_2 terms in equations 11 and 13 become increasingly important as the polymer MW increases, and A_2 can no longer be assumed to be zero. The calculated MW decreases with increasing dn/dc, whereas the size remains unchanged.

MWs and RMS radii of Perspex and UV 52E obtained from ThFFF–MALLS–RI are summarized in Table III. The values of dn/dc and A_2 were taken as 0.083 and 2 × 10^{-4}, respectively, for both polymers. Figures 8 and 9 show the plots of MWD and the RMS radius versus

Table II.   Molecular Weights and RMS Radii of PERSPEX Determined by ThFFF–MALLS–RI Using Different dn/dc and A_2 Values

dn/dc	A_2	M_n	M_w	M_z	$r_{g,n}$ (nm)	$r_{g,w}$ (nm)	$r_{g,w}$ (nm)
0.083	0	4.31 × 10^6	5.87 × 10^6	8.01 × 10^6	91.6	103.8	114.0
0.083	2 × 10^{-4}	4.42 × 10^6	6.10 × 10^6	8.31 × 10^6	92.8	106.0	116.5
0.083	5 × 10^{-4}	4.61 × 10^6	6.55 × 10^6	8.90 × 10^6	95.5	110.7	122.2
0.083	2 × 10^{-4}	4.42 × 10^6	6.10 × 10^6	8.31 × 10^6	92.8	106.0	116.5
0.088	2 × 10^{-4}	4.15 × 10^6	5.73 × 10^6	7.80 × 10^6	92.7	105.7	116.2

Table III.   MW and RMS Radii Determined
by ThFFF–MALLS–RI for PMMA Materials

	M_n	M_w	M_z	$r_{g,n}$ (nm)	$r_{g,w}$ (nm)	$r_{g,w}$ (nm)
Perspex	4.42×10^6	6.10×10^6	8.31×10^6	92.8	106.0	116.5
UV 52E	1.83×10^6	3.77×10^6	8.00×10^6	62.2	84.5	113.6

MW, respectively. In Figure 9, no significant difference in the slope was found between two materials: 0.346 for Perspex and 0.354 for UV 52E. Those slopes are lower than that obtained for the PS standard in Figure 5, which indicates both PMMA materials have higher molecular density than the PS standard.

The PMMA materials were also characterized using conventional ThFFF and SEC without the use of light-scattering detector. RI traces from SEC are shown in Figure 10, and the MWs determined by ThFFF and SEC are summarized in Table IV. For the same sample, the MWs obtained from ThFFF are higher than those obtained from SEC. As the polymer size approaches the exclusion limit of the columns, the MW tends to be underestimated in SEC. There is also a possibility of shear degradation as these ultrahigh MW polymers pass through the SEC columns.

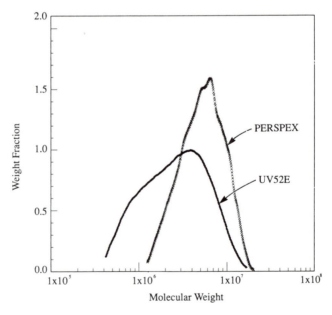

Figure 8.   MW distributions of PMMA materials shown in Figure 6.

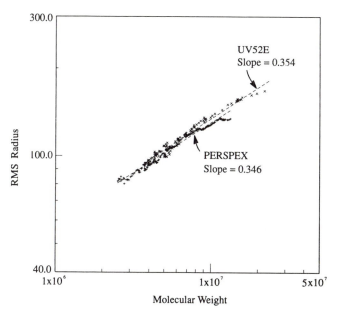

Figure 9.   RMS radius versus MW for PMMA materials shown in Figure 6.

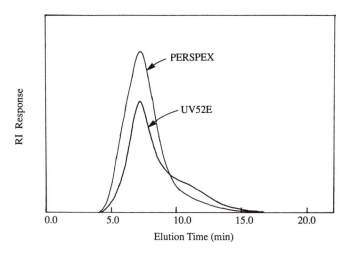

Figure 10.   SEC of PMMA materials in THF at 1 mL/min.

Table IV. MW Determined by ThFFF–RI
and SEC–RI for PMMA Materials

	M_n	M_w	M_z
ThFFF–RI			
Perspex	2.77×10^6	7.11×10^6	11.6×10^6
UV 52E	1.37×10^6	3.99×10^6	7.24×10^6
SEC–RI			
Perspex	0.832×10^6	2.14×10^6	3.58×10^6
UV 52E	0.577×10^6	1.67×10^6	3.06×10^6

Conclusions

With its multiangle capability, MALLS can be used to measure the size as well as the absolute MW of polymers. The multiangle capability seems to be particularly important for the determination of MW and RMS radius of ultrahigh MW polymers as the Debye plot deviates from the linearity. It is also important for the analysis of ultrahigh MW polymers to use accurate A_2 value as well as dn/dc as the resulting MW and RMS radius tend to vary with those values.

Even with an absolute MW detector such as light scattering, the accuracy of the polymer MW determined by a separation technique depends on the resolution of the separator because the MW is calculated based on the assumption that each data slice is monodisperse. Thus, it is important to choose a separation technique that provides higher resolution for the polymers to be analyzed. ThFFF offers higher resolution than SEC for high MW polymers, particularly for polymers with MW near or higher than 1 million Da.

Acknowledgments

We acknowledge Lena Nilsson of Wyatt Technology for helpful discussions on the light-scattering applications.

References

1. Apple, D. J.; Kincaid, M. C.; Mamalis, N.; Olson, R. J. *Intraocular Lenses—Evolution, Design, Complications, and Pathology;* Williams and Wilkins: Baltimore, MD, 1989.
2. Bailey, D.; Vogel, O. *J. Macromol. Sci. Rev. Macromol. Chem.* 1976, *C-14,* 267.
3. Tirrel, D. A. *Polym. News* 1989, *7,* 104.
4. Giddings, J. C. *Chem. Eng. News* 1988, *October 10,* 34.
5. Gunderson, J. J.; Giddings, J. C. *Anal. Chim. Acta* 1986, *189,* 1.
6. Gao, Y. S.; Caldwell, K. D.; Myers, M. N.; Giddings, J. C. *Macromolecules* 1985, *18,* 1272.

7. Giddings, J. C.; Li, S.; Williams, P. S.; Schimpf, M. E. *Makromol. Chem.,*
    Rapid Commun. **1988**, *9,* 817.
8. Lee, S. In *Chromatography of Polymers: Characterization by SEC and FFF;*
    Provder, T. Ed.; ACS Symposium Series 521; American Chemical Society:
    Washington, DC, 1993; pp 77–88.
9. Thompson, G. H.; Myers, M. N.; Giddings, J. C. *Anal. Chem.* **1969,** *41,*
    1219.
10. Hovingh, M. E.; Thompson, G. H.; Giddings, J. C. *Anal. Chem.* **1970,** *42,*
    195.
11. Rudin, A.; Johnston, H. K. *Polym. Lett.* **1971,** *9,* 55.
12. Schimpf, M. E.; Giddings, J. C. *J. Polym. Sci. Polym. Phys. Ed.* **1989,** *27B,*
    1317.
13. Schimpf, M. E.; Giddings, J. C. *J. Polym. Sci. Polym. Phys. Ed.* **1990,** *28B,*
    2673.
14. Kirkland, J. J.; Rementer, S. W. *Anal. Chem.* **1992,** *64,* 904.
15. Martin, M.; Hes, J. *Sep. Sci. Technol.* **1984,** *19,* 685.
16. Wyatt, P. J.; Jackson, C.; Wyatt, G. K. *Am. Lab.* **1988,** *May,* 86.
17. Johann, C.; Kilz, P. *J. Appl. Polym. Sci. Appl. Polym. Symp.* **1991,** *48,* 111.
18. Wyatt, P. J. *J. Liq. Chromatogr.* **1991,** *14(12),* 2351.
19. Gunderson, J. J.; Caldwell, K. D.; Giddings, J. C. *Sep. Sci. Technol.* **1984,**
    19(10), 667.
20. Zimm, B. H. *J. Chem. Phys.* **1948,** *16(12),* 1093.
21. *Polymer Handbook,* 3rd ed.; Brandrup, J.; Immergut, E. H., Eds.; Wiley:
    New York, 1989.
22. Williams, P. S.; Giddings, J. C. *Anal. Chem.* **1987,** *59,* 2038.

RECEIVED for review January 6, 1994. ACCEPTED revised manuscript June 13,
1994.

Molecular Characterization Using a Unified Refractive Index–Light-Scattering Intensity Detector

Robyn Frank,[1] Lothar Frank,[1] and Norman C. Ford[*,2]

[1] Lark Enterprises, 12 Wellington Street, Webster, MA 01570
[2] Precision Detectors, Inc., 160 Old Farm Road, Amherst, MA 01002

We develop series expansions useful in extrapolating two-angle light-scattering data to 0° using the Debye expression for the form factor of a Gaussian coil. Errors that would be encountered if the equations were used to analyze data on molecules of other shapes are discussed. Graphs show the percent error in radius of gyration (R_g) and molecular weight (M_w) for hard spheres, rigid rods, and flexible rings over the range of R_g = 0–150 nm. Finally, we present experimental data showing that instrument calibration done in one solvent can be used in other solvents with different index of refraction and that accurate values of dn/dc can be obtained using a commercially available refractive index detector.

LIGHT-SCATTERING INSTRUMENTS designed to be used on a routine basis as detectors in chromatography systems are commercially available. The instruments measure scattered intensities at two or three angles and are used in conjunction with a concentration detector (often a differential refractometer) to determine the molecular weight (M_w) distribution and, for larger molecules, radius of gyration (R_g). A number of questions arise in obtaining optimum results from these detectors: (1) How should the calculations of M_w and R_g be done?; (2) Over what range of M_w and R_g are the results reliable?; and (3) Does the calibration extend to other solvents and polymers?

Mourey and Coll (*1*) suggested that the form factor for a Gaussian coil be used to analyze two-angle data at 15° and 90° to obtain M_w and

* Corresponding author

R_g. They used an iterative approach to make this calculation and showed that good agreement with the expected values for polystyrene standards in tetrahydrofuran (THF) were obtained for M_w from 1.06×10^3 to 2.3×10^6 Da. They also obtained values for R_g in good agreement with literature values over the range of ~12 to ~72 nm.

First, we provide series expansions that accurately allow the calculation of M_w and R_g using the method of Mourey and Coll without requiring an iterative approach. Second, we present experimental data showing that a calibration using a single standard in one solvent can be used to make measurements in a variety of solvents. We do this by studying two polystyrene standards dissolved in five different solvents with specific refractive index (RI) increments (dn/dc) ranging from 0.0615 to 0.224.

Series Expansions for 1/P(θ) and R_g

The usual starting point for discussion of light-scattering intensity (2) is the equation

$$\frac{i_o}{I_o} = \frac{4\pi^2 n^2 V_o (dn/dc)^2 \sin^2 \vartheta c M_w P(\theta)}{\mathcal{N}\lambda^4 r^2} = S_0 P(\theta) \tag{1}$$

where i_o is the intensity of light in the scattered field, I_o the intensity of incident light, n the solvent index of refraction, V_o the illuminated sample volume, ϑ the angle between polarization direction and scattering direction, c the concentration, $\mathcal{N}$ Avogadro's number, λ_o the light wavelength in vacuum, and r the distance from the scattering volume to the detector. $P(\theta)$ is the form factor depending on the scattering angle, θ. It is equal to 1.0 for molecules much smaller than λ_o/n and decreases with increasing molecular size. S_0 is the intensity that would be obtained if the measurement were made at a scattering angle of 0°. We have specialized equation 1 for linearly polarized light and for values of c sufficiently low that virial coefficients may be neglected. The light-scattering geometry is shown in Figure 1.

Modern light-scattering photometers use a small diameter laser beam as a light source. In this case, we can replace $I_o V_o$ with $P_o l$, where P_o is the laser power output and l the path length of the laser beam that actually contributes to the detected signal. Calculation of the detector response also requires integration over both the area of the detector and the length of the laser beam contributing to the detector signal. In practice, it is more convenient to calibrate the instrument using a single well-characterized narrow distribution standard rather than attempting calibration from a knowledge of the scattering geometry.

The instrument used in these studies measures light scattered at two angles. Figure 2 shows the optical arrangement of the detector. Using

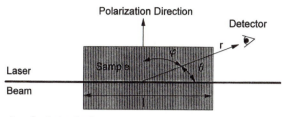

- θ = Scattering Angle
- φ = Angle between Polarization & Scattering Directions
- r = Distance from Sample to Detector
- l = Length of Scattering Region

Figure 1.   Diagram showing the light scattering geometry.

a Fourier lens optical system, light scattered from 14° to 16° is collected in a single detector. A second detector collects light scattered at 90° using a flat-ended GRIN lens 2 mm in diameter purchased from NSG America, Inc., Somerset, NJ. The maximum acceptance angle (δ_m) of the GRIN lens is limited by its numerical aperture to 27.5° in the center of the lens when it is in contact with a fluid having an index of refraction $n = 1.0$. The acceptance angle for other fluids is given approximately by $\delta_m \cong 27.5°/n$. The acceptance angle is also reduced for light entering the lens at positions nearer its edge. The detector solid angles are ∼0.057 steradians at 15° and 0.15 steradians at 90° for a fluid with $n = 1.4$.

The detector solid angles depend on the index of refraction of the fluid in the cell because of refraction at the fluid–window interface. The solid angles are proportional to $1/n^2$ for both detectors. However, the

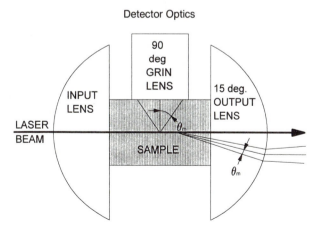

Figure 2.   Optical configuration of the two-angle light-scattering photometer used in these studies.

length of the laser beam contributing to the signal is limited by apertures for the 15° detector so that $1 \propto n$, whereas the 90° path length is not dependent on n. Consequently, with the factor of n^2 in equation 1, we expect the detected signal to be independent of n at 90° and proportional to n at 15°.

The danger associated with a finite acceptance range is that averaging $P(\theta)$ over the full range may give a result for $P(\theta)$ different from that obtained if the angular range is very small. We begin a discussion of this problem by using the asymptotic expression for $P(\theta)$,

$$P(\theta) = 1 - (\tfrac{1}{3})(4\pi n/\lambda_o)^2 R_g^2 \sin^2 (\theta/2) = 1 - f \sin^2 (\theta/2) \qquad (2)$$

where $f = (\tfrac{1}{3})(4\pi n/\lambda_o)^2 R_g^2$ is introduced to reduce the complexity of some of the equations to follow. The detected signal (LS_θ) is obtained by substituting equation 2 into equation 1 and integrating over the solid angle of the detector.

$$LS_\theta = I_o S_0 \int \{1 - f \sin^2 (\theta/2)\} d\Omega \qquad (3)$$

Where $d\Omega$ is an increment of the detector solid angle. For the 15° detector we can use the approximation $\sin (\theta/2) \approx \theta/2$ and $d\Omega = 2\pi\theta d\theta$. With these substitutions, equation 3 may be written

$$LS_{15} = 2\pi I_o S_0 \int_{\theta - \delta_m}^{\theta + \delta_m} (1 - f\theta^2/4)\theta d\theta = 4\pi\theta\delta I_o S_0(1 - (\theta^2 + \delta^2)/4) \qquad (4)$$

Thus, the detected signal obtained for collection over the angular range of $\theta - \delta_m$ to $\theta + \delta_m$ is equivalent to a signal obtained at a scattering angle θ_{eff} with

$$\theta_{eff}^2 = \theta^2 + \delta^2. \qquad (5)$$

If $\theta = 15°$ and $\delta_m = 1°$, $\theta_{eff} = 15.03°$, indistinguishable in a practical sense from 15°.

At 90°, integration of equation 2 from $90 - \delta_m$ to $90 + \delta_m$ shows that $\theta_{eff} = 90°$ to all orders in δ for any collection function symmetric about 90°. However, equation 2 will become increasingly inaccurate as R_g increases beyond 40 nm. We need therefore to consider higher order terms. They depend on the shape of the molecule, but we can gain some insight into the potential of a serious error by considering the well-known Debye expression for Gaussian coils (3),

$$P(\theta) = 2u^{-2}(e^{-u} + u - 1)$$
$$u = (4\pi n \sin (\theta/2)/\lambda_o)^2 R_g^2 = 3 f \sin^2 \theta/2 \qquad (6)$$

Assuming the lens collects all light in a cone with a half angle of δ_m, the total light collected by the 90° detector will be

$$LS_{90} = I_oS_0 \int_{-\delta_m}^{\delta_m} P(\pi/2 + \delta)(\delta_m{}^2 - \delta^2)^{1/2}d\delta \qquad (7)$$

We have evaluated equation 7 using numerical techniques. We used values of $n = 1.4$ and $\lambda_o = 670$ nm in equation 6, and did the calculation for $R_g = 20$–140 nm and $\delta_m = 10$–$30°$. Surprisingly, the ratio

($P(\theta)$ averaged over the detector solid angle)/$P(90)$

did not differ from 1 by >3% for $\delta_m = 20°$ over the range in R_g studied. A summary of the results is given in Table I. We conclude that it is valid to treat the 90° light-scattering signal without correcting for the finite collection aperture.

The detected signal is given by

$$LS_{90} = (4\pi^2P_0l_{90}\Omega_{90} \sin^2 \vartheta/\mathcal{N}\lambda_o{}^4)(dn/dc)^2cM_wP(90)$$
$$= K_{90}(dn/dc)^2cM_wP(90) \qquad (8a)$$
$$LS_{15} = (4\pi^2P_0l_{15}\Omega_{15} \sin^2 \vartheta/\mathcal{N}\lambda_o{}^4)n(dn/dc)^2cM_wP(15)$$
$$= K_{15}n(dn/dc)^2cM_wP(15) \qquad (8b)$$

The constants K_{15} and K_{90} are independent of both solvent and solute. They can in principle be determined in a single measurement of a narrow distribution standard with $R_g < 10$ nm. Notice that the light-scattering signal amplitudes are proportional to the constants K_{15} and K_{90}. To increase the signal strength, and therefore, the signal-to-noise ratio, the instrument designer must increase these factors. The key quantity is

**Table I.   Errors in 90° Scattering
Intensity Due to Finite Detector
Acceptance Angle**

δ_m	R_g	$P(90)$	$P(\theta)avg/P(90)$
10°	50	0.86822	1.00011
10°	100	0.60195	1.00121
10°	150	0.38011	1.00336
20°	50	0.86822	1.00043
20°	100	0.60195	1.00477
20°	150	0.38011	1.01343
30°	50	0.86822	1.00094
30°	100	0.60195	1.01051
30°	150	0.38011	1.03009

$P_o l \Omega / \lambda^4$. It is this product rather than any single term, such as the laser power, that determines the total signal detected.

A third calibration constant K_{RI} obtained at the same time as K_{15} and K_{90} gives the relationship between the concentration and RI signal,

$$RI = K_{RI}(dn/dc)c \qquad (9)$$

K_{RI} can be obtained under the assumptions that no sample is retained by the column and that a single value of dn/dc describes the entire sample. Integration of equation 9 over the full RI curve gives the area under the curve, A_{RI}.

$$A_{RI}Q = K_{RI}(dn/dc)M_I \qquad (10)$$

where M_I is the mass of sample injected onto the column and Q is the mobile phase flow rate. Equation 10 can be used to determine either K_{RI} or dn/dc if the other is known.

We now turn to a discussion of the development of equations that can be used to obtain M_w and R_g from the three quantities, RI, LS_{15}, and LS_{90}. We seek series expansions for the quantities $1/P(15)$ and nR_g that agree reasonably with equation 6. The goal is to obtain agreement within 0.1% for $R_g < 140$ nm.

We have tried two simple series expansions in terms of the ratio:

$$R = \frac{LS_{15}}{LS_{90}} = \frac{P(15)}{P(90)} \qquad (11)$$

They are first,

$$1/P(15) = 1 + \Sigma a_i(R - 1)^i$$
$$nR_g = \Sigma b_i(R - 1)^{i/2} \qquad (12)$$

and second,

$$1/P(15) = 1 + \Sigma c_i(1 - 1/R)^i$$
$$nR_g = \Sigma d_i(1 - 1/R)^{i/2} \qquad (13)$$

The first choice was found to converge to the desired accuracy with fewer terms than the second and is used in the remainder of this development.

We first obtain a_1 and b_1 by substituting equation 2 into equation 11. For $\lambda_o = 670$ nm we find

$$(nR_g)^2 = \frac{R - 1}{5.6632 \times 10^{-5} + 5.8629 \times 10^{-5}(R - 1)} \qquad (14)$$

or, as $R - 1 \rightarrow 0$

$$nR_g \approx 132.882(R - 1)^{1/2}$$

$$1/P(15) \approx 1 + 0.03527(R - 1) \tag{15}$$

Using the coefficients in equation 15 as the first terms in the series expansions of equation 12 to assure that the slopes of the curves as $R - 1 \rightarrow 0$ are correct and obtaining additional coefficients using the program given by Bevington (4) to fit data to a polynomial, we obtain with all weights set equal to 1,

$$nR_g = 132.882(R - 1)^{1/2} + 0.218(R - 1) - 18.907(R - 1)^{3/2}$$
$$+ 11.574(R - 1)^2 - 1.886(R - 1)^{5/2}$$
$$1/P(15) = 1 + 0.03527(R - 1) - 0.0063303(R - 1)^2$$
$$+ 0.0015949(R - 1)^3 \tag{16}$$

Expressions for $P(\theta)$ have been given for a variety of molecular shapes (5). Figures 3 and 4 show the errors that would be encountered if data for solid spheres, flexible rings, or rigid rods were analyzed using equation 12. (The molecular weight of a solid sphere with $R_g = 40$ nm would be in excess of 5×10^8 Da, so the larger errors for solid spheres at $R_g > 40$ nm are not relevant for gel permeation chromatography (GPC) applications.) We conclude that light-scattering measurement at 15° and 90° will give M_w with a "shape" error of <1% for molecules with $R_g < 100$ nm (<80 nm for hard spheres) and R_g with a "shape" error <10% with $R_g < 100$ nm (<65 nm for hard spheres) if the data are analyzed using equation 16.

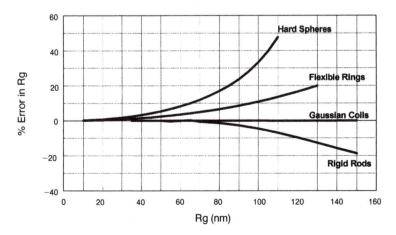

Figure 3. Errors expected in R_g using equation 12 to calculate R_g for molecules of different shape.

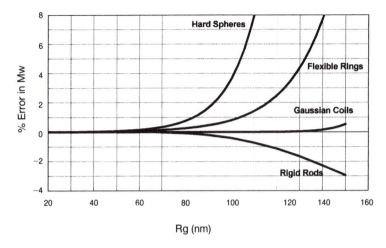

Figure 4.   Errors expected in using equation 12 to calculate M_w for molecules of different shape.

Series expansions similar to equations 16 can also be obtained for other molecular shapes for which $P(\theta)$ is known and for other pairs of scattering angle. Table II gives the coefficients in equations 12 for several shapes.

The scattering angles were 90° and 15° in all cases. The accuracy of the fits over the range of R_g = 0–100 nm and R_g = 0–150 nm are also given. These coefficients may be used to give more accurate data analysis in cases where the molecular shape is known.

Experimental Study of Effects of Solvent Index of Refraction

We have examined experimentally the response of the light-scattering instrument to scattering using the same polystyrene standards in solvents of different index of refraction. We made a series of measurements on two polystyrene standards (Millipore nominal M_w = 8500, P/N 25171, Lot #80314 and Waters nominal M_w = 110,000, P/N 41995, Lot #70111) dissolved in five solvents [acetone, methyl ethyl ketone (MEK), tetrahydronaphthalene, THF, and toluene].

The two standards, supplied to us by Millipore and Waters, were manufactured and characterized by Pressure Chemical (Pittsburgh, PA). They provide characterization data for the nominal 8500 Da standard showing M_w = 8000, M_n = 7350, and M_p = 8640 by size-exclusion chromatography and M_n = 9050 by vapor pressure osmometry. For the purposes of this study, we based all calculations on a value of 8500 daltons for this standard. No errors were quoted for this standard.

The nominal 110,000 Da standard was listed as having M_w = 93,050 ± 6% by light scattering, M_v = 98,700 ± 6% by intrinsic viscosity, and M_n

= 92,600 ± 6% by membrane osmometry. The molecular weight calculated from the stoichiometry of the polymerization reaction was 100,000 ± 6% Da.

Samples were prepared with ~30 mg of standard in 3 mL solvent for the 8500 Da standard and 5.5 mg of standard in 3 mL solvent for the 110,000 Da standard. The concentration of the samples was estimated to be accurate to ±1.5%. Injections of 100 µL were used in all cases. Five to 10 runs were made for each sample with the exception of the 110,000 M_w sample in acetone (the sample did not dissolve).

Chromatography was done using a Waters model 510 pump, a Waters model 712 WISP injector, and a single Toyasoda TSK300W silicon gel column. The same column was used for all solvents; it was flushed for ≥12 h with a new solvent before making measurements.

A Precision Detectors model PD2000 light-scattering photometer (scattering angles of 15° and 90°) was mounted in a Waters model 410 refractometer. The two detectors were connected in series with the refractometer last to avoid damage to the refractometer due to excessive pressure. (The photometer can withstand pressures in excess of 1000 psi.) It was determined in a separate experiment using human serum albumin in a salt buffer that the effective interdetector volume was 90 ± 10 µL. The photometer cell volume was 10 µL. Data were collected at 1-s intervals in all cases.

Data Analysis

Of a total of 60 runs, one was discarded because of a steeply sloping RI baseline and one was discarded because of an injector failure. The analysis of the remaining runs is the subject of this section.

The first step was the determination of the three calibration constants using equations 8a, 8b, and 10. We used the nominal 8500 Da sample dissolved in THF for calibration purposes, as this sample can be expected to have a form factor very nearly equal to 1.0 even at a scattering angle of 90°. We also assumed that dn/dc = 0.180 obtained by extrapolating literature data (6) to the wavelength of the laser in our instrument, 670 nm.

We next used equation 10 to estimate dn/dc for polystyrene in each of the other solvents. A comparison of our results with literature values is given in Table III. The excellent agreement shows that an ordinary differential refractometer can be used to obtain values of dn/dc for a solvent-solute pair with an unknown dn/dc.

The light source used in the refractometer had a central wavelength of 930 nm. Dispersion in the index of refraction of polystyrene between the RI wavelength and LS wavelength (670 nm) is expected to have no effect in this study because the same sample was used in all solvents. However, measurements made in other materials could show effects of the optical dispersion on the accuracy of the dn/dc determinations. Consequently, for general laboratory use, it is desirable to change the light source in the refractometer to more nearly match the laser wavelength.

Table II. Coefficients for Use in Equation 12 for Several Molecular Configurations

	Gaussian Coil	Hard Sphere	Rigid Rod	Flexible Ring
a_1	0.03527	0.03527	0.03527	0.03527
a_2	-0.0063303	-0.016492	-0.021729	-0.0098190
a_3	0.0015949	0.0039449	0.015746	0.0015076
$R_g < 100$ nm	< ±0.07%	< ±0.4%	< ±0.4%	< ±0.2%
$R_g < 150$ nm	< ±0.50%	Not useful above 100 nm	< ±2.0%	< ±2.0%
b_1	132.882	132.882	132.882	132.882
b_2	0.2178	0.3173	-0.5031	0.2699
b_3	-18.907	-44.144	-12.665	-30.935
b_4	11.574	21.450	-7.870	16.187
b_5	-1.886	-2.886	15.565	-2.596
$R_g < 100$ nm	< ±0.05%	< ±0.3%	< ±0.22%	< ±0.08%
$R_g < 150$ nm	< ±0.11%	Not useful above 100 nm	< ±0.4%	< ±0.10%

Table III. Values of dn/dc Compared with Literature Values

Solvent	dn/dc (This Research)	dn/dc (Literature)	λ (Literature) (nm)
Acetone	0.224 ± 0.004	0.240 (Extrapolated from acetone cyclohexane mixtures)	546
MEK	0.213 ± 0.004	0.214–0.230	546
Tetrahydronaphthalene	0.0615 ± 0.002	0.072–0.075	436
THF	0.180 (Assumed value)	0.186–0.193	632
Toluene	0.108 ± 0.002	0.104–0.125	546

Finally, we calculated $M_wP(15)$ and $M_wP(90)$ for each sample using equations 8a, 8b, and 9. In principle, we could calculate these values for each 1-s slice, construct a molecular weight distribution, and then calculate a weighted average. However, we note that for a narrow standard, $P(\theta)$ is constant across the elution peak so

$$P(\theta)M_w = \frac{\Sigma P(\theta)_i M_i c_i}{\Sigma c_i} \tag{17}$$

where the sum is over all slices. Because $P(\theta)_i M_i = LS_i/c_i$ we have

$$P(\theta)M_w = \frac{\Sigma LS_i}{\Sigma c_i} = \frac{\text{area under } LS \text{ curve}}{\text{area under } RI \text{ curve}} \tag{18}$$

(We have suppressed factors of dn/dc, n, and the calibration constants for clarity.) We calculated $P(\theta)M_w$ by first summing all runs for a given sample and then using equation 18. We estimated the standard deviation for each sample by calculating $P(\theta)M_w$ individually for each run and doing a statistical analysis of the group of runs from each sample.

When the signal-to-noise ratio of the light-scattering chromatogram is high, the molecular weights are very consistent from run to run with standard deviations of <1%. We have established an empirical relationship between the standard deviation in M_w determination, σ (in %) and the reciprocal of the signal-to-noise ratio for the chromatogram, $\sigma(\%)$ $\approx 10(S/N)^{-1}$. This means that a signal-to-noise ratio of only 10 will lead to molecular weight measurements accurate to ~1%. The validity of this relationship will depend on the number of slices in each peak. We obtained a measurement each second and the peaks were ~90 s wide. In principle, the accuracy of the measurement could be improved by slowing down the mobile phase flow rate to obtain more points under the peak. However, noise due to particulate matter in the solvent and baseline drift will at some point overcome any advantage obtained by a slower flow rate.

The values of $P(\theta)M_w$ for two samples and at both scattering angles are plotted as a function of the index of refraction of the solvent in Figure 5. This figure shows quite clearly that a single calibration point in one solvent (the nominal 8500 Da standard in THF) can be used to measure molecular weights in other solvents. With the exception of the low molecular weight standard in tetrahydronapthalene (dn/dc = 0.0615), which gave a signal-to-noise ratio <5, all the low molecular weight points were within 3% of 8325 Da. The high molecular weight points were all within 4.6% of 95,485 Da. Both molecular weights are in agreement with the values given by Pressure Chemical given the uncertainties in both the Pressure Chemical values and our light-scattering measurement.

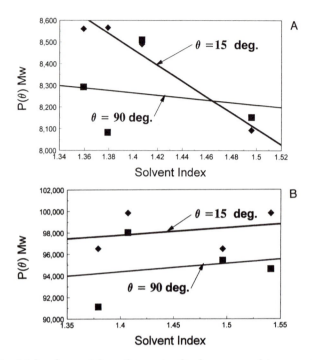

Figure 5. Molecular weights of two standards measured in a number of solvents plotted as a function of the index of refraction of the solvent. ■, data points for 90° measurements; ◆, for 15° measurements. (A) Nominal 8500 Da standard; (B) nominal 110,000 Da standard.

It is important to remember that the uncertainty in measurements includes a possible systematic effect due to an error in dn/dc for one or more of the solvents, as well as a statistical component due to a finite signal-to-noise ratio in the light scattering. It is possible that more accurate values of dn/dc would reduce the discrepancies in values of $P(\theta)M_w$ among the different solvents.

The data on the nominal 110,000 Da sample show that, within the experimental error, the index of refraction dependence of the detected light scattering is as predicted in equations 8a and 8b. Equation 8a predicts that light scattering at 90° will have no direct factors of n because of the competition between a factor of n^2 in the light-scattering equation (eq 1) and a factor of $1/n^2$ in the equation giving the solid angle over which light is collected. Equation 8b, on the other hand, predicts that light scattering at 15° will be proportional to n because the effective laser path length is proportional to n. The fact that the data agree with these predictions is direct evidence that we correctly understand the detector optics.

Conclusions

We give series expansions that may be used to obtain M_w and R_g from light-scattering measurements at two angles, 15° and 90°, together with an RI measurement. Expansions are given for several molecular shapes. It is pointed out that reasonable accuracy is obtained for all molecular shapes studied, and for a useful range of molecular sizes, if the results for a Gaussian coil are used to analyze data for molecules of any of the shapes considered here. Thus, the expressions are useful in implementing the method suggested by Mourey and Coll (*1*) for obtaining M_w and R_g from two-angle light-scattering measurements.

We also show experimentally that a single calibration of the light-scattering instrument can be used for measurements of molecules in solvents other than the calibration solvent. Values of dn/dc were obtained from the RI detector signal. Thus, the light-scattering–RI detector may be used to obtain M_w even in cases where dn/dc is not known. There are, however, two restrictions that apply. The sample may not be retained by the column. Otherwise, calculation of dn/dc will be in error. A single value of the specific RI increment must characterize the entire sample. A mixture of molecules with different values of dn/dc would not be correctly analyzed.

References

1. Mourey, T. H.; Coll, H. Chapter 10 in this volume.
2. Tanford, C. *Physical Chemistry of Macromolecules;* John Wiley & Sons: New York, 1961; Chapter 5.
3. Debye, P. *J. Phys. Colloid Chem.* **1947,** *51*, 18.
4. Bevington, P. R. *Data Reduction and Error Analysis for the Physical Sciences;* McGraw-Hill: New York, 1969; Chapter 8.
5. Kerker, M. *The Scattering of Light and Other Electromagnetic Radiation;* Academic: Orlando, FL, 1969; Chapter 8; Casassa, E. In *Polymer Handbook,* 3rd ed.; Brandrup, J.; Immergut, E. H., Eds.; John Wiley & Sons: New York, 1989; Part VII, p 485.
6. Huglin, M. B. In *Polymer Handbook,* 3rd ed.; Brandrup, J.; Immergut, E. H., Eds.; John Wiley & Sons: New York, 1989; Part VII, p 409.

RECEIVED for review January 6, 1994. ACCEPTED revised manuscript November 7, 1994.

Size-Exclusion Chromatography with Light-Scattering Detection at Two Angles

Polystyrene in Tetrahydrofuran

Thomas H. Mourey and Hans Coll

Analytical Technology Division, Research Laboratories B–82, Eastman Kodak Company, Rochester, NY 14650–2136

A method for the analysis of data from a size-exclusion chromatography (SEC) detector that measures elastic light-scattering intensities at two angles (15° and 90°) is evaluated for linear polystyrenes in tetrahydrofuran (THF). Over certain size ranges a single detector can be used to calculate molecular weights by assuming the particle-scattering function to be unity. The 90° scattering is useful for isotropic scatterers less than ~70,000 MW and the 15° scattering for polystyrenes less than ~500,000 MW. For anisotropic scatterers, the ratio of scattering intensities at the two angles is used to calculate the particle scattering function and root-mean-square radius, assuming a specific polymer shape (e.g., random coil). Scattering at the low angle and the particle-scattering function are then used to calculate weight-average molecular weights. It is shown that the assumption of shape has only a minor effect on the calculation of polymer sizes in the size range fractionated by common SEC columns. Accuracy and precison of measured radii are greatly affected by detector noise on the ratioing method, insensitivity of the light-scattering detector to small molecules in broad polymer distributions, interdetector volume, and data fitting. However, one method that uses the ratio of areas of the light-scattering signals alone calculates with high precision and accuracy an average radius that corresponds most closely to a z-average.

Elastic light-scattering detection for size-exclusion chromatography (SEC) has evolved in two directions: low-angle laser light scattering

0065–2393/95/0247–0123$12.00/0

(LALLS) in which measurements are made at a single angle, typically less than 7° (1, 2); and multiangle laser light scattering (MALLS), which measures scattered light at angles typically between $\sim$15° and 160° (3, 4). The low angle of LALLS approximates the zero-angle scattering intensity, thus simplifying calculation of the weight-average molecular weight, $\overline{M}_w$. MALLS relies on graphical methods to obtain intercepts and limiting (zero-angle) slopes of Debye (or related) plots, resulting in $\overline{M}_w$ and, for large polymers, the z-average of the root-mean-square radius of gyration, $\bar{r}_{gz}$. Both methods have strengths and weaknesses. Simultaneous measurement of light-scattering intensities at two angles (in this case 15° and 90°) is a compromise between the two techniques. It does require, however, a different approach to data analysis than is currently used for LALLS and MALLS detectors. An appreciation for the differences can be gained from an overview of conventional light-scattering data analysis methods for SEC light-scattering detectors.

Theory

Local Properties. "Local properties" are values such as detector response, molecular weight, or polymer size at a particular retention volume of a size-exclusion chromatogram. They are denoted by the subscript i. The excess Rayleigh scattering, $R_{\theta i}$, at each retention volume v_i of an SEC is related to the concentration at each retention volume, c_i, and angle, θ, by

$$\frac{Kc_i}{R_{\theta i}} = \frac{1}{M_{wi}P(\theta)_i} + 2A_{2i}c_i + \cdots \tag{1}$$

$P(\theta)_i$ is the particle scattering function and A_{2i} is the second virial coefficient at each retention volume. M_{wi} is molecular weight at each retention volume and is a weight average if molecules of more than one molecular weight elute at the same retention volume. $R_{\theta i}$ is measured by the light-scattering detector and c_i is obtained from an independent concentration detector such as a differential refractive index (DRI) detector. K is the optical constant for light-scattering intensity perpendicular to the plane of polarized incident light,

$$K = \frac{4\pi^2 n^2 (dn/dc)^2}{\lambda_0^4 N_A} \tag{2}$$

where n is the refractive index of the solvent, dn/dc is the polymer specific refractive index increment, λ_0 is the wavelength of light in vacuum, and N_A is Avogadro's number. For the collection of scattered light through an annular opening, such as in LALLS instruments, the optical constant for plane-polarized incident light is

$$K = \frac{2\pi^2 n^2 (dn/dc)^2 (1 + \cos^2 \theta)}{\lambda_0^4 N_A} \tag{3}$$

K is the same as for unpolarized incident light with a fixed-point detector. The second and higher concentration terms of equation 1 are usually negligible at the low concentrations used in SEC. In this case,

$$M_{wi} \approx \frac{R_{\theta i}}{K c_i P(\theta)_i} \tag{4}$$

At low angles, the particle-scattering function approaches unity even for relatively large particle (polymer) sizes, further simplifying the calculation of molecular weights using equation 4. At higher angles, $P(\theta)$ is commonly given in the form of a power series in $\sin^2 \theta/2$, which for $c_i = 0$ leads to the familiar result used in the reciprocal scattering plots of Zimm (5).

$$\mathscr{L}_{c=0} \frac{K c_i}{R_{\theta i}} = \frac{1}{M_{wi}} \left(1 + \frac{16\pi^2 n^2}{3\lambda_0^2} r_{gi}^2 \sin^2 \frac{\theta}{2} + \cdots \right) \tag{5}$$

Equation 5 can be applied in the analysis of MALLS data. Most important, this solution for $P(\theta)$ becomes independent of particle shape as θ approaches zero. In practice, scattering intensities are collected at a number of angles. The value of the limiting slope of the Zimm plot at $c = 0$ is proportional to the square of the radius of gyration (z-average), and the intercept is $1/M_{wi}$.

Scattering intensities at only two angles are not suited for the graphical methods based on equation 5, unless in a limited size range where the higher terms of the particle-scattering function are insignificant. One alternative is to measure the ratio of scattering intensities, Z_i, at angles θ_1 and θ_2, which is equal to the ratio of particle-scattering functions at these two angles,

$$Z_i = \frac{P(\theta_1)_i}{P(\theta_2)_i} = \frac{R_{\theta_1 i}}{R_{\theta_2 i}} \tag{6}$$

The ratio of two angles symmetric about 90°, commonly known as a "disymmetry" measurement, has been used for many years to measure polymer and particle sizes (6); however, the two angles need not be symmetric about 90° for equation 6 to apply. Sizes can be calculated from this ratio using an analytical relationships for $P(\theta)$ for a specific particle shape. The relationship given by Debye (7) for random coils is commonly used for polymer molecules,

$$P(\theta) = \frac{2}{x^2} [e^{-x} + x - 1] \qquad (7)$$

where

$$x = \frac{16\pi^2 n^2}{\lambda_0^2} r_g^2 \sin^2 \frac{\theta}{2} \qquad (8)$$

and n is the solvent refractive index and λ_0 is the wavelength of light in vacuum.

Equation 7 applies to monodisperse linear Gaussian chains. The solution for spheres (8, 9) is given by

$$P(\theta) = \left[\frac{3}{x^3} (\sin x - x \cos x)\right]^2 \qquad (9)$$

where

$$x = 4\pi\left(\frac{Rn}{\lambda_0}\right) \sin \frac{\theta}{2} \qquad (10)$$

and

$$R = r_g \sqrt{\frac{5}{3}} \qquad (11)$$

This expression is useful for globular polymers of radius R. Also, the spherical model is the form that is approached as branching in a polymer molecule increases (6).

Analysis of light-scattering data at two angles begins with the calculation of excess Rayleigh scattering at each retention volume for the two light-scattering chromatograms, from which is calculated the ratio, Z_i, of 15° to 90° scattering. If the ratio is 1.0 then the molecules are isotropic scatterers, and either the 15° or 90° light-scattering data can be used to calculate molecular weights by using $P(\theta) = 1.0$; however, no size information is obtained. If the scattering at 15° is greater than the scattering at 90°, then

1. Assume a shape and substitute the appropriate particle-scattering function into equation 6. For a random coil, the expression is

$$Z_i = \frac{\dfrac{2}{x_{\theta 1}{}^2}\left[e^{-x_{\theta_1}} + x_{\theta_1} - 1\right]}{\dfrac{2}{x_{\theta 2}{}^2}\left[e^{-x_{\theta_2}} + x_{\theta_2} - 1\right]} \tag{12}$$

2. Find the value of r_g by iteration that satisfies equation 12, where x is given by equation 8. This value is r_{gi}, the root-mean-square radius at retention volume v_i.

3. Calculate $P(\theta)_i$ from this value of r_{gi} using equations 7 and 8.

4. Use $P(\theta)_i$ and equation 1 (or equation 2) to calculate M_{wi}. In this study, M_{wi} is calculated from $P(15)_i$ and $R_{15}i$.

Average Properties. "Average properties" of the entire polymer are calculated from the local values measured in the SEC light-scattering experiment. We define molecular weight averages as follows:

Number-average molecular weight:

$$\overline{M}_n = \frac{\sum c_i}{\sum \dfrac{c_i}{M_{wi}}} \tag{13}$$

Weight-average molecular weight:

$$\overline{M}_w = \frac{\sum c_i M_{wi}}{\sum c_i} \tag{14}$$

Z-average molecular weight:

$$\overline{M}_z = \frac{\sum c_i M_{wi}{}^2}{\sum c_i M_{wi}} \tag{15}$$

The whole polymer $\overline{M}_w$ can also be obtained without the DRI, from the sample mass injected, m, and the volume increment between data points, Δv_i.

$$\overline{M}_w = \frac{1}{Km} \sum \frac{R_{\theta i}}{P(\theta)_i} \Delta v_i \tag{16}$$

The weight-average molecular weight is of particular significance in this study because it can be compared with values measured by static LALLS on unfractionated polymer.

Similar equations are used to calculate average sizes for the whole polymer:

$$\bar{r}_{gn} = \left[\frac{\Sigma\left(\frac{c_i}{M_{wi}}\right)r_{gi}^{\,2}}{\Sigma\left(\frac{c_i}{M_{wi}}\right)} \right]^{1/2} \tag{17}$$

$$\bar{r}_{gw} = \left[\frac{\Sigma c_i r_{gi}^{\,2}}{\Sigma c_i} \right]^{1/2} \tag{18}$$

$$\bar{r}_{gz} = \left[\frac{\Sigma c_i M_{wi} r_{gi}^{\,2}}{\Sigma c_i M_{wi}} \right]^{1/2} \tag{19}$$

Substituting equation 4 into equation 19 gives an equation analogous to equation 14 for the whole polymer z-average radius of gyration, r_{gz}, that requires quantities obtained from the light-scattering detector alone (no DRI).

$$\bar{r}_{gz} = \left[\frac{\Sigma \frac{R_{\theta i}}{P(\theta)_i} r_{gi}^{\,2}}{\Sigma \frac{R_{\theta i}}{P(\theta)_i}} \right]^{1/2} \tag{20}$$

Methods that calculate average polymer properties without the DRI are significant because they circumvent complications that arise from the measurement of the interdetector volume between the light-scattering and concentration detectors (10, 11). It is necessary, however, that each local signal, $R_{\theta i}$, be divided by the computed particle scattering function, $P(\theta)_i$.

The ratio of the areas under the light-scattering chromatograms at two angles,

$$\bar{Z} = \frac{\Sigma R_{15i}}{\Sigma R_{90i}} \tag{21}$$

can be used to measure a new undefined root-mean-square radius of gyration. The ratio is used in equation 12 to solve by iteration for a value of the root-mean-square radius for the whole polymer that we call $\bar{r}_{gu}$. This is again of interest because it is obtained from the light-scattering detector alone. In addition, the areas of the light-scattering chromatograms can be measured with high precision and accuracy, making this a simple method for obtaining an average polymer size.

Experimental Details

A PD2000W light-scattering detector (Precision Detectors, Amherst, MA) was installed in a Waters Corporation (Milford, MA) model 410 DRI immediately before the DRI cell. The light source is a plane-polarized solid-state laser emitting at 670 nm. As described by the manufacturer, the instrument measures 15° scattered light through an annular opening with a solid angle of $\sim$0.06 sr and at a fixed point at 90° with a scattering solid angle of $\sim$0.8 sr. Data were collected at a sampling rate of 4 points per second using a 16-bit Data Translation (Marlborough, MA) 2805–5716 A/D board. Data acquisition and analysis software was written in-house in ASYST 4.0.

Three 7.5 mm i.d. × 300 mm PLgel 5 μm Mixed-C columns (Polymer Laboratories, Amherst, MA) were thermostated to 30.0°C. Uninhibited tetrahydrofuran (THF) at a nominal flowrate of 1.0 mL/min was used as the eluent. Narrow distribution polystyrene standards (Polymer Laboratories) were injected in a volume of 100 μL with concentrations ranging from 0.1 to 2.5 mg/mL, depending on molecular weight. Broad molecular-weight-distribution polystyrene #18,242-7 was obtained from Aldrich Chemical Company (Milwaukee, WI). Acetone, added to each sample at a concentration of 0.1%, was used as a flow marker.

Specific refractive index increments of polystyrene standards were measured at 632.8 nm in a Thermo Separations Products (Riviera Beach, FL) KMX-16 differential refractometer at 30.0 °C. Specific refractive index increments at 670 nm were estimated by extrapolation of the data at 632.8 nm and published values at 436 and 546 nm (6) for polystyrene with molecular weight 500,000,

$$(dn/dc)_{670} \approx 0.9783(dn/dc)_{632.8}$$

The molecular weight dependence of polystyrene dn/dc at 670 nm was obtained from the following:

$$(dn/dc)_{670 \, nm} = 0.1804 - 9.149/M$$

A specific refractive index increment of 0.180 at 670 nm was used for the broad polystyrene sample. All dn/dc values were used in light-scattering calculations at three significant figures. Calibration factors to convert voltage output to R_θ for the 15° and 90° light-scattering detectors were calculated using equation 4 for $P(\theta) = 1.0$ from the response of multiple injections of PS 26,700, PS 19,400, and PS 22,000. The $\overline{M}_w$ values of each standard were confirmed by static LALLS, and the calibration factors measured for these three standards were averaged together.

$\overline{M}_w$ values were measured on a KMX-6 LALLS (Thermo Separations Products) at 6–7° in THF using a 15-mm static LALLS cell and a 0.15-mm aperture.

Results and Discussion

Polystyrene molecules with molecular weights less than $\sim$100,000 are isotropic scatterers at the wavelength of light used by this instrument. Identical excess Rayleigh scattering should be observed at 15° and 90°, and data from either angle can be used to calculate molecular weights using equation 4 with $P(\theta) = 1.0$; however, no size information is obtained. The same molecular weights are measured at the two angles

Table I.   Low-Molecular-Weight
            Polystyrenes

Vendor $\overline{M_w}$	$\overline{M_w}$ 15°	$\overline{M_w}$ 90°
61,600	64,600	60,300
52,000	48,600	45,200
26,700	26,400	25,700
19,400	21,300	20,000
9200	9300	9150
7000	7260	7130
5050	5290	5230
3250	3130	3090
2450	2700	2630
1700	ND	1770
1060	ND	1060

ND, not determined, noisy light-scattering signal.

(Table I), although best results are obtained using the 90° scattering
data for the lowest molecular weight samples because this detector is
less sensitive to particulates in the eluent.

   Light-scattering chromatograms of a narrow polystyrene standard
with a peak molecular weight reported by the vendor of 1,030,000 are
shown in Figure 1. Now, greater scattering intensity is clearly observed

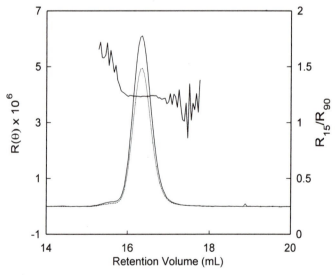

Figure 1.   Light-scattering chromatograms at 15° (———) and 90°
(· · · · ·) of narrow polystyrene with M_p = 1,030,000 and the ratio of the
two curves.

at the lower angle. The ratio of these curves, also shown in Figure 1, exhibits the least amount of scatter in the middle where both light-scattering signals are strong. Values of r_{gi} for narrow polystyrene standards were calculated using equation 12 from the ratios of excess Rayleigh scattering at each retention volume, Z_i. These values of r_{gi} were then used to calculate $P(15)_i$ and finally M_{wi} with equations 7 and 4, respectively. $\overline{M}_w$ values measured by SEC light-scattering (column 3 of Table II) of narrow distribution polystyrenes with molecular weights >100,000 agree with values measured by static LALLS to within 4%. Values of $\overline{M}_w$ calculated from the 15° light-scattering detector assuming $P(\theta) \approx 1$ are shown in column 4 of Table I. The assumption of $P(\theta) \approx 1$ has no effect on values of $\overline{M}_w$ for polystyrenes less than ~500,000 MW, and the error introduced is ~1.3% at a molecular weight of 1,030,000 and ~2.2% at molecular weight 2,300,000.

Assuming a random coil shape, the undefined radius, $\bar{r}_{gu}$, obtained from the ratio of peak areas of the excess Rayleigh chromatograms, is comparable to $\bar{r}_{gz}$ values calculated from the light-scattering and DRI signals and equation 19. Precision of $\bar{r}_{gu}$ and $\bar{r}_{gz}$ values improves as the ratio of excess Rayleigh scattering at 15° and 90° for the whole polymer ($\overline{Z}$ values in column 5 of Table II) increases. $\overline{Z}$ is measured to better than ±1% for all samples. One result is that $\bar{r}_{gu}$, measured from only a ratio of areas of excess Rayleigh scattering chromatograms, is the simplest and most precise average size measured, particularly at small sizes.

Both $\bar{r}_{gu}$ and $\bar{r}_{gz}$ values in Table II are in reasonable agreement with values calculated from the literature (*12*). The following power laws (for $\bar{r}_g$ units of nanometers) are obtained from least squares fits:

$$\bar{r}_{gz} = (0.0166 \pm 0.0009)\overline{M}_w^{0.57+0.02} \tag{23}$$

$$\bar{r}_{gu} = (0.0121 \pm 0.0005)\overline{M}_w^{0.60+0.02} \tag{24}$$

Both are in agreement, within experimental error, with the theoretical prediction of an exponent of ~0.58 (*13*). A similar scaling relationship was also presented in a preliminary study with this instrument (*14*). This previous study used an earlier version of light-scattering detector flow cell on polystyrene standards that were of similar but not identical molecular weight.

Particle-scattering functions for random coils and spheres are indistinguishable at values of $\overline{Z}$ near unity (Figure 2). Likewise, the radii of random coils and spheres are also similar in the range of Z values less than ~1.2 (Figure 3). This is reflected in the similarity of radii calculated for random coils and spheres (columns 7 and 8 of Table II). The largest difference between radii calculated for these two shapes is 8.5% for

Table II. High-Molecular-Weight Polystyrenes

| Vendor | Static LALLS $\overline{M}_w$ | Equation 4 | | | Equation 13 | | | |
		Random Coil $\overline{M}_w$	$P(\theta)=1$ $\overline{M}_w$	$\overline{Z}$	Random Coil $\bar{r}_{gz}$ (nm)	Random Coil $\bar{r}_{gu}$ (nm)	Sphere $\bar{r}_{gu}$ (nm)	Lit.[a] $\bar{r}_{gz}$ (nm)
2,300,000	2,280,000	2,240,000 ± 38,000[b] (1.7%)[c]	2,190,000	1.641 ± 0.007 (0.4%)	71.8 ± 0.3 (0.4%)	72.1 ± 0.4 (0.6%)	65.4	75.2
1,260,000	1,260,000	1,240,000 ± 27,000 (2.2%)	1,160,000	1.302 ± 0.010 (0.8%)	50.4 ± 0.7 (1.4%)	50.5 ± 0.9 (1.8%)	48.0	53.1
1,030,000	995,000	1,030,000 ± 16,000 (1.2%)	1,017,000	1.237 ± 0.005 (0.4%)	44.5 ± 4.5 (10.4%)	44.5 ± 0.8 (1.8%)	43.3	47.6
775,000	679,000	684,000 ± 12,000 (1.8%)	680,000	1.156 ± 0.008 (0.7%)	36.9 ± 0.9 (2.4%)	36.9 ± 0.8 (2.2%)	36.3	37.4
570,000	528,000	524,000 ± 3,700 (0.7%)	522,000	1.113 ± 0.005 (0.4%)	31.7 ± 1.0 (3.2%)	32.0 ± 0.6 (1.9%)	31.4	32.0
336,000	325,000	312,000 ± 1,100 (0.4%)	312,000	1.062 ± 0.003 (0.3%)	22.9 ± 0.9 (3.9%)	23.6 ± 0.7 (3.0%)	22.7	23.6
155,000	149,000	152,000 ± 900 (0.6%)	152,000	1.030 ± 0.005 (0.5%)	16.8 ± 5.3 (31.5%)	15.0 ± 1.6 (10.7%)	14.3	15.5
127,000	125,000	124,000 ± 1,900 (1.5%)	124,000	1.022 ± 0.003 (0.3%)	12.5 ± 1.8 (14.4%)	12.2 ± 1.3 (10.7%)	11.7	13.7

[a] Lit. are literature values calculated from reference 13.
[b] Sample estimate of the standard deviation, s for six samples.
[c] Coefficient of variation, defined as (s/mean) × 100.

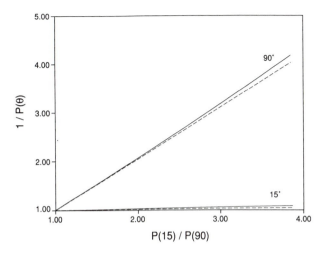

Figure 2. Reciprocal of particle-scattering functions for random coils (————) and spheres (– – –).

polystyrene 2,300,000. This is at least encouraging for the application of this method to branched polymers; the most extreme radial density distribution of an unswollen branched polymer is that of a solid sphere, which is uncommon even for compact structures such as dendrimers (16). It is expected that for most branched systems, the random coil particle-scattering function is more suitable.

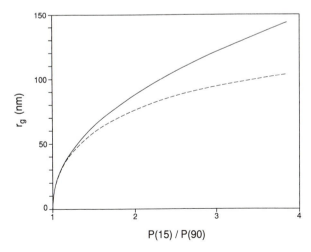

Figure 3. Values of r_g as a function of $P(15)/P(90)$ for solvent refractive index 1.405. Random coils (————), spheres (– – –).

Log M–retention volume calibration data constructed from the narrow polystyrene standards in Table I are shown in Figure 4. Also shown is a calibration curve for a broad polystyrene sample (solid line) calculated from the 15° light-scattering detector and the DRI, after accounting for the volume delay (interdetector volume) between the two detectors. The two sets of calibration data are similar; however, a substantial portion of the molecular weight distribution of the broad sample was not measured because the light-scattering detector is insensitive to the small molecules in the distribution. This point has been addressed in light-scattering and viscometry detection through linear extrapolation of log M or log $[\eta]$ versus retention volume plots in the region of the concentration distribution for which there is no measurable light-scattering signal. Weighting factors used in the extrapolation have been derived based on standard error propagation equations (17). The dashed line in Figure 4 represents this extrapolated region and provides an estimate of molecular weight over nearly half of the polymer distribution defined by the concentration chromatogram. As is well known, the number-average molecular weight is most affected by the small molecules in the distribution, and $\overline{M}_n$ is biased high without this extrapolation.

Extrapolations can also be applied to log r_g-retention volume plots (dashed line in Figure 5) to provide a better estimate of the number-average radius. There is reasonable superposition of the r_g values of the broad standard on the narrow standard calibration data, but unlike the log M plot in Figure 4, there is considerable scatter in the data at long retention volumes (small sizes). In this region the 15° and 90° signals are nearly identical and the ratio of the two signals depends heavily on

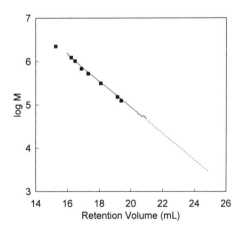

Figure 4.   Log M–retention volume data for narrow polystyrene standards (■) and a broad molecular weight distribution polysyrene. – – –, the extrapolated region.

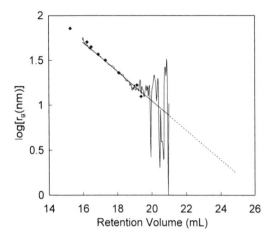

Figure 5.   *Log* r_g*–retention volume data for narrow polystyrene standards* (■) *and a broad molecular weight distribution polystyrene. ———, the extrapolated region; – – –, fifth-order fit to the raw data.*

the amount of detector noise. The scatter in r_{gi} values is translated through to plots of log M versus log r_g, which are often used to deduce polymer conformation. In Figure 6, there is curvature and only fair superposition of the data for the broad standard on data for narrow standards. An average slope of 0.60 ± 0.13 (sample standard deviation) obtained from 20 chromatograms of broad polystyrene is consistent with a random coil polymer in a good solvent, but the uncertainty in the slope is far too large to rule out other conformations.

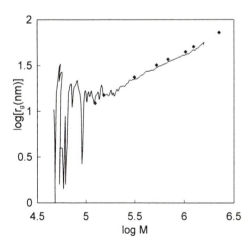

Figure 6.   *Log M versus local root-mean-square radius for a broad poly-styrene standard (———) and narrow polystyrene standards (■).*

The choice of interdetector volume affects both the slope and shape of plots such as in Figure 6. In this study, interdetector volume was obtained from the volume required to superimpose the DRI chromatogram on a light-scattering chromatogram of a narrow-distribution polystyrene standard. The use of "effective" interdetector volumes (18) that also partially correct for axial dispersion can lead to slopes for plots such as in Figure 6 of <0.5, which is clearly incorrect for a random coil polymer. This effect of interdetector volume on conformation plots has been reported by others (19) and indicates a potential danger in deducing conformations without full appreciation of the factors that affect the accuracy and precision of these plots.

Average properties for the broad polystyrene standard calculated from local values that were extrapolated in the low-molecular-weight region are given in Table III. The precision of molecular weight averages is comparable or slightly better than the precision reported for LALLS and MALLS instruments operated in THF at room temperature (19), and the $\overline{M}_w$ agrees with the value measured by static LALLS. Much of the data are extrapolated at small sizes (Figure 4), contributing to the poorer reproducibility of $\overline{M}_n$. In comparison with molecular weight averages, average radii listed under "unfitted $\bar{r}_{gi}$ and $P(\theta)_i$" are inconsistent with each other [e.g., $\bar{r}_{gn} > \bar{r}_{gw}$ and $\bar{r}_{gz} \neq \bar{r}_{gz}$ (no DRI)] and imprecise. This is caused by the uncertainty in r_{gi} at small sizes (scatter in the data at long retention volume is Figure 5), and in the case of $\bar{r}_{gz}$ (no DRI)

Table III.   Aldrich 18,242-7 Polystyrene

$\overline{M}_n$	$134,000 \pm 7800^a$ $(5.8\%)^b$
$\overline{M}_w$	$303,000 \pm 1000$ (0.3%)
$\overline{M}_z$	$490,000 \pm 3900$ (0.8%)
$\overline{M}_w$ (no DRI, eq 16)	$301,000 \pm 1700$ (0.6%)
$\overline{M}_w$ (static LALLS)	$298,000 \pm 2000$ (0.7%)
Unfitted r_{gi} and $P(\theta)_i$	
$\bar{r}_{gn}$	29.8 ± 31.8 (106.7%)
$\bar{r}_{gw}$	29.0 ± 16.6 (57.2%)
$\bar{r}_{gz}$	30.1 ± 4.3 (14.3%)
$\bar{r}_{gz}$ (no DRI, eq 20)	52.8 ± 22.1 (41.9%)
$\bar{r}_{gu}$ (no DRI, eq 21)	28.7 ± 0.8 (2.8%)
Fitted r_{gi} and $P(\theta)_i$	
$\bar{r}_{gn}$	14.3 ± 1.2 (8.4%)
$\bar{r}_{gw}$	22.2 ± 0.8 (3.6%)
$\bar{r}_{gz}$	28.7 ± 0.5 (1.7%)
$\bar{r}_{gz}$ (no DRI, eq 20)	26.1 ± 2.1 (8.0%)
$\bar{r}_{gu}$ (no DRI, eq 21)	28.7 ± 0.8 (2.8%)

[a] Sample estimate of the standard deviation, s, for 20 samples.
[b] Coefficient of variation, defined as (s/mean) × 100.

scatter in the measured particle-scattering function (Figure 7). Averages in Table III listed under "fitted r_{gi} and $P(\theta)_i$" were calculated from the fitted data in Figures 5 and 7. This fitting has no effect on the values or precision of molecular weight averages. It has a significant effect on average radii; reasonable values are obtained and precision is greatly improved. The undefined root-mean-square radius of gyration, $\bar{r}_{gu}$, corresponds most closely to the z-average radius of gyration. The z-average radius obtained without the DRI is within experimental error the same as the $\bar{r}_{gz}$ values obtained from the light-scattering and DRI detector, but the large sample standard deviation indicates sensitivity to the data fitting procedure applied.

Concerns about factors that affect the particle-scattering function can be raised about dissymmetry methods when applied to whole polymers (6). For example, the Debye function does not account for excluded volume effects in thermodynamically good solvents, which can perturb the Gaussian behavior of linear chains. Particle-scattering functions that account for excluded volume effects (20) could be more suitable for some polymer-solvent systems. Modifications can also be made to the particle-scattering function for polydispersity, although they are small in the size range examined in this study (6). More important, SEC provides fractionation according to size and each slice of a chromatogram is more nearly monodisperse than is the whole polymer.

Conclusions

The molecular weight of polystyrene less than ~500,000 MW can be measured accurately in THF with a two-angle light-scattering instrument

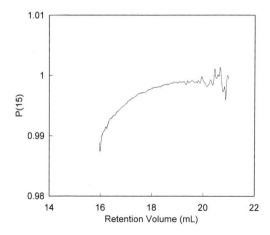

Figure 7.   Particle-scattering function at 15° measured for broad polystyrene standard (——) and fifth-order fit constrained to pass through 1.0 at longest retention volume (- - -).

operating at 670 nm by using the scattering data at 15° and assuming $P(\theta) = 1$. Light-scattering data at 90° can be used in the same manner for isotropic scatterers, which in this instrument corresponds to polystyrenes with molecular weights less than $\sim$70,000. The molecular weight of polymers >500,000 MW can be measured accurately from the 15° scattering by calculating the particle-scattering function for a specific shape (random coil) from the ratio of the light-scattering chromatograms at 15° and 90°. The root-mean-square radii of polystyrenes that are anisotropic scatterers may also be obtained from the ratio of light-scattering chromatograms at 15° and 90°, although a shape must be assumed. Fortunately, the choice of the particle shape has no effect on the calculation of molecular weights and only a small effect on the calculation of radii for the range of sizes fractionated by many SEC columns.

The average radius obtained from the ratio of excess Rayleigh chromatograms corresponds to the z-average root-mean-square radius and is obtained with high precision and requires no knowledge of interdetector volume. In comparison, conformation plots and the calculation of other average radii are greatly affected by noise, insensitivity of the light-scattering detector to small molecules, and the value of interdetector volume used in calculations. Rational, precise results can be obtained for number- and weight-average radii, but only through data fitting. Although demonstrated for data at 15° and 90°, the method may also be used with two other angles, including the classic 45° and 135° configuration used in dissymmetry measurements, with corrections by the appropriate particle-scattering functions. Some utility may be found in using this method as a check of the graphical procedures used with MALLS (more than two angles) detectors.

Glossary of Symbols

A_2	second viral coefficient
c_i	concentration at retention volume v_i
dn/dc	specific refractive index increment
K	light-scattering optical constant
m	sample mass injected
M_{wi}	local weight-average molecular weight at retention volume v_i
$\overline{M}_n$	number-average molecular weight of the whole polymer
M_p	SEC peak molecular weight
$\overline{M}_w$	weight-average molecular weight of the whole polymer
$\overline{M}_z$	z-average molecular weight of the whole polymer
n	refractive index of solvent
N_A	Avogadro's number

$P(\theta)$ particle-scattering function at angle θ
$P(\theta)_i$ particle-scattering function at angle θ and retention volume v_i
$P(15)$ particle-scattering function at $15°$
$P(90)$ particle-scattering function at $90°$
R particle radius
R_{15} excess Rayleigh scattering at $15°$
R_{90} excess Rayleigh scattering at $90°$
$R_{\theta i}$ excess Rayleigh scattering at angle θ and retention volume v_i
R_{15i} $15°$ excess Rayleigh at retention volume v_i
R_{90i} $90°$ excess Rayleigh scattering at retention volume v_i
r_g root-mean-square radius of gyration
r_{gi} local z-average root-mean-square radius of gyration at retention
         volume v_i
$\bar{r}_{gn}$ number-average root-mean-square radius of gyration of the
         whole polymer
$\bar{r}_{gw}$ weight-average root-mean-square radius of gyration of the whole
         polymer
$\bar{r}_{gz}$ z-average root-mean-square radius of gyration of the whole
         polymer
S variance
v retention volume
w_i weighting factor at retention volume v_i
x particle-scattering function variable defined by equations 7 and 9
Z_i ratio of excess Rayleigh scattering at 15 and $90°$ at retention
         volume v_i
$\bar{Z}$ ratio of excess Rayleigh scattering at 15 and $90°$ for the whole
         polymer
Δv_i volume increment between data points
θ angle of observation of scattered light
λ_0 wavelength of light in vacuum

Acknowledgments

Thanks are extended to Catherine Harrison and Brian Owens for their expert laboratory assistance and to Norman Ford of Precision Detectors, Inc. for assistance and enlightening discussions on the light-scattering instrument used in this study.

References

1. Ouano, A. C. *J. Polym. Sci., Part A-1* **1972**, *10*, 2169.
2. Kaye, W. *Anal. Chem.* **1973**, *45*, 221A.
3. Wyatt, P. J.; Jackson, C.; Wyatt, G. K. *Am. Lab.* **1988**, *20*, 86.
4. Wyatt, P. J.; Hicks, D. L.; Jackson, C.; Wyatt, G. K. *Am. Lab.* **1988**, *20*, 108.
5. Zimm, B. H. *J. Chem. Phys.* **1948**, *16*, 1099.

6. *Light Scattering from Polymer Solutions;* Huglin, M. B., Ed.; Academic: Orlando, FL, 1972.
7. Debye, P. *J. Phys. Colloid Chem.* **1947,** *51,* 18.
8. Lord Rayleigh *Proc. Roy. Soc. (London)* **1911,** *A84,* 25.
9. Gans, R. *Ann. Phys. (Leipzig)* **1925,** *76,* 29.
10. Mourey, T. H.; Miller, S. M. *J. Liq. Chromatogr.* **1990,** *13,* 693.
11. Cheung, P.; Balke, S. T.; Mourey, T. H. *J. Liq. Chromatogr.* **1992,** *15,* 39.
12. Schulz, V. G. V.; Bauman, H. *Makromol. Chem.* **1968,** *114,* 122.
13. LeGuillou, J. C.; Zinn, J. *J. Phys. Rev. Lett.* **1977,** *39,* 95.
14. Mourey, T. H.; Coll, H. *Polym. Mater. Sci. Eng.* **1993,** *69,* 217.
15. Mourey, T. H.; Turner, S. R.; Rubinstein, M.; Fréchet, J.; Hawker, C.; Wooley, K. *Macromolecules* **1992,** *25,* 2410.
16. Lew, R.; Cheung, P.; Balke, S. T.; Mourey, T. H. *J. Appl. Polym. Sci.* **1993,** *47,* 1685.
17. Mourey, T. H.; Balke, S. T. In *Chromatography of Polymers: Characterization by SEC and FFF;* Provder, T., Ed.; ACS Symposium Series 521; American Chemical Society: Washington, DC, 1993, pp 180–198.
18. Shortt, D. W. ; Wyatt, P. J. *International GPC Symposium Proceedings;* Waters Corporation: Milford, MA, 1994; p 121.
19. Jeng, L.; Balke, S. T.; Mourey, T. H.; Wheeler, L.; Romeo, P. *J. Appl. Polym. Sci.* **1993,** *49,* 1359.
20. Ptitsyn, O. B. *Zh. Fiz. Khim.* **1957,** *31,* 1091.

RECEIVED for review January 6, 1994. ACCEPTED revised manuscript July 6, 1994.

Characterization by Size-Exclusion Chromatography with Refractive Index and Viscometry
Cellulose, Starch, and Plant Cell Wall Polymers

Judy D. Timpa†

Agriculture Research Service, Southern Regional Research Center, U.S. Department of Agriculture, New Orleans, LA 70179

High molecular weight natural polymers are difficult to characterize because isolation, extraction, and dissolution often degrade the polymers and no good molecular weight standards exist. In our laboratory, cellulose, starch, and plant cell wall materials have been directly dissolved in the nondegrading solvent dimethylacetamide–lithium chloride (DMAC–LiCl) without prior isolation or extraction. Size-exclusion chromatography with viscometry and refractive index detectors was used with DMAC–LiCl as the mobile phase. The universal calibration concept was applied to obtain molecular weight distributions (MWDs). Applications include cotton fiber, corn and wheat starch flours, and avocado cell walls. Relationships were determined between the respective MWDs and cotton fiber variety, inheritance, textile processing, and strength; starch extrusion conditions; and avocado ripening stage.

N ATURAL POLYMERS SUCH AS POLYSACCHARIDES, which usually have high molecular weight (MW) components, are difficult to characterize. Appropriate analytical techniques are generally dependent on getting the polymer into solution. Isolation and extraction often alter the polymer composition (1). Available solvents have serious limitations, most often because of degradation. MW standards are not generally available. The solvent dimethylacetamide–lithium chloride (DMAC–LiCl) offers the capacity for a wide a range of applications for dissolution of cellulose, starch, chitin, and other polysaccharides with little or no degradation

† Deceased.

(2–4). Cotton fibers are single cells composed primarily (~96%) of the polymer cellulose. In our laboratory (5), cotton fibers were dissolved directly in the solvent DMAC–LiCl. This procedure solubilizes fiber cell wall components directly without prior extraction or derivatization, processes that could lead to degradation of high MW components. MW determinations have been carried out by a size-exclusion chromatography (SEC) system using commercial columns and instrumentation with DMAC–LiCl as the mobile phase. Incorporation of viscometry and refractive index (RI) detectors (6) allowed application of the universal calibration concept (7) to obtain MW distributions (MWDs) based on well-characterized narrow-distribution polystyrene standards (5). The universal calibration concept used by incorporation of dual detectors bypasses the need for cellulose standards. There are no cellulose standards available. Polystyrene standards for a wide range of MWs dissolved readily in DMAC–0.5% LiCl with no activation necessary.

We extended the methodology developed for molecular characterization of cotton fiber to analysis of other polysaccharides. In this report, we present the results obtained from MWDs determined by SEC for various complex carbohydrate samples dissolved in DMAC–LiCl. Applications include cotton fiber, corn and wheat starch flours, and avocado cell walls. Relationships are evaluated between the respective MWDs and cotton fiber development, variety, inheritance, textile processing, and strength; starch extrusion conditions; and stage of ripening in avocado.

Experimental Details

Safety Considerations. N,N-dimethylacetamide is an exceptional contact hazard that may be harmful if inhaled or absorbed through the skin and may be fatal to embryonic life in pregnant females (Baker Chemical Company, N,N-dimethylacetamide, Material Safety Data Sheet, 1985, D5784–01; pp 1–4).

Sample Preparation. Samples were dissolved as previously described (5). Ground material was added to DMAC (Burdick & Jackson, Muskegon, IL) in a Reacti-Vial (Pierce, Rockford, IL) in a heating block. Activation was achieved by elevating the temperature to 150 °C and maintained at that temperature for 1–2 h. The temperature was lowered to 100 °C followed by addition of dried LiCl (~8% wt/vol). Samples were held at 50 °C until dissolved (18–48 h) and subsequently were diluted and filtered. Final concentration of samples was 0.9–1.2 mg/mL in DMAC with 0.5% LiCl. At least two dissolutions per sample were made for subsequent SEC analysis.

Chromatography. Filtered polysaccharide solutions were analyzed using an SEC system consisting of an automatic sampler (Waters WISP, Waters, Milford, MA) with a high-performance liquid chromatography pump (Waters model 590), pulse dampener (Viscotek, Houston, TX), viscometer detector (Viscotek model 100), and RI detector (Waters model 410). The detectors were connected in series. The mobile phase was DMAC–0.5% LiCl pumped at a flow rate of 1.0 mL/min. Columns were Ultrastyragel 10^3, 10^4, 10^5, 10^6 (Waters) preceded by a guard column (Phenogel, linear, Phenomenex, Torrance, CA). A column heater (Waters Column Temperature System) regulated the temperature of the columns at 80 °C. Injection volume was 400 μL with a run time of 65 min. The software package Unical based on ASYST (Unical, Version 3.02, Viscotek) was used for data acquisition and analysis. Calibration was with polystyrene standards ranging in MW from 6.2 × 10^3 to 2.9 × 10^6 (Toyo Soda Manufacturing, Tokyo, Japan) dissolved and run in DMAC–0.5% LiCl. The universal calibration curve was a logarithmic function of the product of the intrinsic viscosity times MW versus retention volume with a third-order fit shown in Figure 1.

Results and Discussion

Dissolution of Cotton Cellulose.   Attempts to identify the true MW of native cellulose always lead to difficulties, especially in isolating unchanged celluloses from natural plant products and in determining the MW of high MW celluloses by reliable physical methods (*1*). The Updegraff procedure is a frequently used method for measuring the

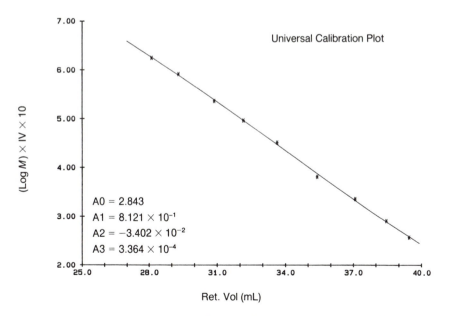

Figure 1.   Universal calibration plot of polystyrene standards dissolved in DMAC–LiCl. (Log M) × IV is hydrodynamic volume, where M is molecular weight and IV is intrinsic viscosity.

cellulose content of plant material (8). Noncellulosic material is removed by treatment with acetic acid–nitric acid reagent at elevated temperature. The remaining cellulose is hydrolyzed to glucose by treatment with concentrated sulfuric acid. Glucose content of the sample is then determined colorimetrically. Often, only the first step of the procedure is used to remove the noncellulosic material. We investigated the effects of the acetic–nitric reagent on the cellulosic composition. The MWD of cotton fiber exposed to acetic–nitric reagent was compared with the native fiber sample. As indicated in Table I, the effect of acid treatment was a shift of the MWD to a lower MW range corresponding to a 93% reduction in MW (9). After acetic–nitric reagent treatment, there is no evidence of low MW cellulose corresponding to polymers found in the primary wall. This suggests that cellulose found in the primary cell wall of cotton fiber is very susceptible to hydrolysis by the Updegraff reagent. Separate treatments of native cotton with acid produced similar average values for MW, but the differences in the broadness of the distribution of cellulosic chains indicates that the extent of polymer degradation is not exactly reproducible. Thus, direct solubilization of cotton fiber cell wall components in DMAC–LiCl without prior extraction or derivatization avoids degradation of the polymer chains and is the preferred method for MWD determinations of cotton fiber cellulose.

Applications. *Monitoring Cotton Fiber Development.* Cotton fiber develops according to specific stages with formation of a primary wall followed by deposition of a secondary layer containing most of the cellulose (10, 11). The biochemical composition of the fiber cell walls is changing throughout development; monitoring those changes has limited progress in this research area (12, 13). Cotton fiber begins development on the day flowering (anthesis); thus, the age of the fiber is usually designated by days after flowering or days postanthesis (DPA). The primary wall stage designates the rapid elongation of the outer cell wall occurring up to ∼21 DPA, whereas the secondary wall development subsequently occurs with the major cellulose deposition. Mature fiber

Table I.   The Effects of Acid Treatment on the MW of Native Cotton Fiber

Sample	$DP_W{}^a$	$DP_N{}^b$	Polydispersity
Native cotton fiber (*Gossypium hirsutum* L.) Texas Marker-I	13,800	4900	2.8
After treatment with acetic–nitric reagent	1050	170	6.2

^{a} DP_W is weight average of polymerization.
^{b} DP_N is number average of polymerization.

is generally harvested from plants at ~60 DPA. In our studies using SEC with dual detectors (*14*), we observed that cell wall polymers from fibers at primary cell wall stages had lower MWs than the cellulose from fibers at the secondary wall stages, as shown in Figure 2. High MW cellulose characteristic of mature cotton was detected as early as 8 DPA. High MW material decreased during the period of 10–18 DPA with concomitant increase in lower MW wall components, possibly indicating hydrolysis during the later stages of elongation.

Monitoring Cotton Fiber Inheritance of Molecular Properties. The relationship between the MWDs, inheritance, and strength of cotton fiber in different genotypes was investigated (*15, 16*). Fiber samples from a new higher (~10%) strength variety for the Mississippi Delta (Mississippi Delta 51, MD51 ne) were assayed and compared with a popular commercial variety and recurrent parent (Deltapine 90). The F1 cross and two selected backcrossed progenies were evaluated from three replications grown in the field. The fiber samples had very similar genetic backgrounds and physical properties except for strength. SEC with viscometry and RI was used to determine the MWD of cotton fiber samples dissolved directly in DMAC–LiCl. As shown in Figure 3, samples of the higher strength variety had a greater proportion of higher MW material than the commercial variety. The most significant difference was apparent in the MW range of 1,000,000–10,000,000.

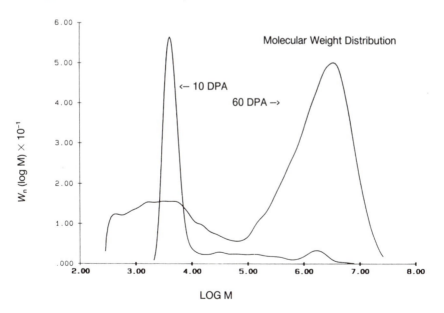

Figure 2. MWDs of samples of cotton fiber at different stages of development: 10 DPA or primary wall stage versus 60 DPA or mature fiber.

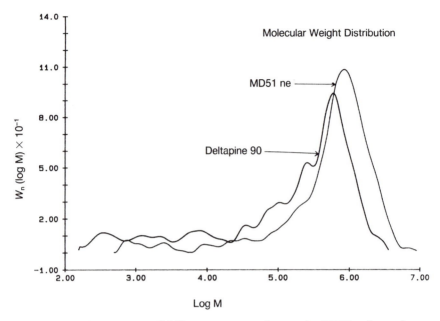

Figure 3.   Comparison of different varieties of cotton by MWDs of samples
of mature cotton fiber. Mississippi Delta 51 (MD 51 ne) has ~10% higher
fiber strength than Deltapine 90.

Effects of Textile Processing.   Mercerization has been used to impart
desirable qualities to cotton fiber for many years. Changes in crystalline
structure of cotton cellulose are one of the major results. In our laboratory (17), alterations in the molecular composition of cotton fiber
caused by mercerization was measured by SEC. Samples from Deltapine,
Acala, and Pima varieties were evaluated as scoured or as mercerized
fiber. SEC determinations of the MWDs of cotton fiber samples before
and after caustic mercerization showed significant loss in the higher
MW fractions for each of the three varieties. Comparison of MWDs for
the acala cotton sample and corresponding mercerized treatment are
provided in Figure 4. Mercerization by liquid ammonia also affected
the higher MW components, although changes in molecular composition
were different from that observed with caustic mercerization.

Effects of Extrusion on Starch.   Processing starch by extrusion results in molecular fragmentation. The effects on the MWDs of flours
from wheat and corn starch were determined (18–20). Starch flours varied in amylose, amylopectin, and protein content. Samples were subjected to twin-screw extrusion with varying moisture content, screw
speed, die temperature, mass flow rate, and protein content. Starch flours
were directly dissolved in the solvent DMAC–LiCl without prior iso-

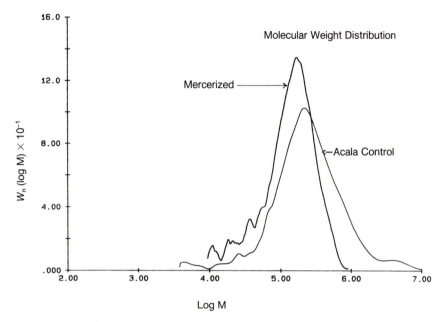

Figure 4.   The effect of NaOH mercerization on cotton fiber of the acala variety.

lation or extraction. SEC using DMAC–LiCl as the mobile phase was used to monitor fragmentation using the same configuration–system developed for cotton fiber studies. Significant extrusion-induced reductions in high MW fractions were observed compared with the native controls as shown in Table II for high amylopectin corn flours. An example of the effect of extrusion on the high protein wheat flour is given in the comparison in Figure 5. The interaction of moisture and die temperature had significant effects on the fragmentation patterns. Qualitative as-

Table II.   The Effect of Extrusion on High Amylopectin Corn Flour (20% moisture, wt/wt): Weight Average Molecular Weights and Relative Abundance of a Molecule Falling within Specific MW Ranges

Sample	Screw Speed (rpm)	Die Temperature (°C)	$M_w \times 10^6$	MW $\geq 10^7$ (%)	MW $10^7\text{--}10^6$ (%)
Control			18.8	26	26
G31	250	140	7.4	13	27
G32	500	140	6.0	8	29
G33	500	180	7.3	17	26
G34	250	180	9.4	14	26

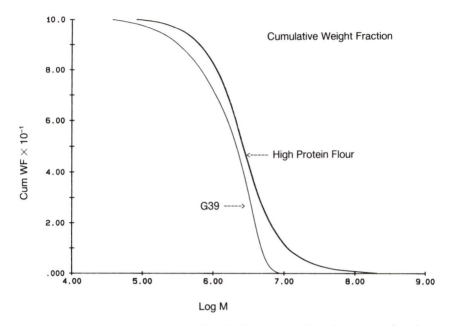

Figure 5.   Cumulative MWDs of the high protein wheat flour control and
an extruded sample (G39).

sessments of branching by monitoring the intrinsic viscosity have been
carried out. For example, shifts in Mark–Houwink plots are evident for
branched samples versus linear types (Figure 6).

Monitoring Fruit Ripening and Cell-Wall Turnover.   Avocados are
a popular fruit for market. The desirability depends on the softening of
the flesh of the fruit. Our research identified key components in the
composition of the avocado cell wall and the changes that occur during
stages of ripening (21). The cellulose molecular structure and crystalline
association of cell walls during fruit ripening was monitored and related
to the levels of the enzyme cellulase. Cellulase is an enzyme that specifically degrades cellulose. SEC techniques used to study cotton fiber
molecular structure were used to evaluate avocado cellulose, and X-ray
diffraction and electron microscopy were used to look at the cellulose
fibers. SEC of total cell wall polysaccharides (including cellulose) revealed a slight increase in the fraction of the largest polymers during
ripening, whereas the crystallinity index increased. Based on these results, we propose that the cellulase prefers to attach the noncrystalline
portions of the cellulose in the wall. This mode of action affects the
firmness of the avocado fruit during the ripening process.

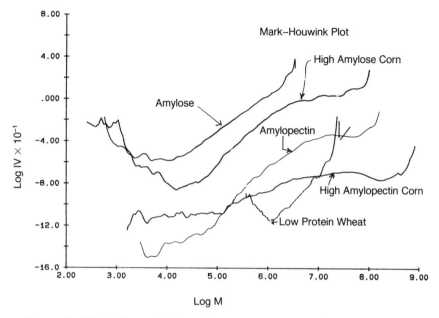

Figure 6.   Mark–Houwink plots of intrinsic viscosities of starch standards (amylose and amylopectin) compared with wheat and corn flours.

Summary

Dissolution of whole plant cell walls or commercially important complex carbohydrates directly into DMAC–LiCl was advantageous. Molecular characterization by SEC with viscometry and RI detectors using the universal calibration allowed evaluation not readily attainable previously (22). Determination of the MWDs of cell wall polymers at critical stages of development provides a tool for understanding biological regulation of the growth processes in cotton fiber and avocado. In addition, monitoring effects of commercial processing of natural polymers assists in minimizing losses and improved end-use products.

Acknowledgments

I thank the following people for their cooperation: D. J. Huber, W. R. Meredith, E. M. O'Donoghue, M. Politz, H. H. Ramey, B. A. Triplett, B. P. Wasserman, A. Striegel, and S. H. Zeronian.

References

1. Franz, G.; Blaschek, W. In *Methods in Plant Biochemistry;* Dey, P. M.; Harbrone, J., Eds.; Academic: Orlando, FL, 1990; Vol. 2, pp 291–322.

2. Turbak, A. B. In *Wood and Agricultural Residues;* Soltes, E.J., Ed.; Academic: Orlando, FL, 1983; pp 87–99.
3. McCormick, C. L.; Callais, P. A.; Hutchinson, B. H. *Macromolecules* **1985,** *18,* 2394–2401.
4. Dawsey, T. R.; McCormick, C. L. *Rev. Macromol. Chem. Phys.* **1990,** *C30,* 403–440.
5. Timpa, J. D. *J. Agri. Food Chem.* **1991,** *39,* 270–275.
6. Haney, M. A. *Am. Lab.* **1985,** *17,* 116–126.
7. Grubisic, A.; Rempp, P.; Benoit, H. A. *Polym. Lett.* **1967,** *5,* 753–759.
8. Updegraff, D. M. *Anal. Biochem.* **1969,** *32,* 420–424.
9. Timpa, J. D.; Triplett, B. A. *Plant Physiol. (Life Sci. Adv.)* **1992,** *11,* 253–256.
10. Marx-Figini, M. In *Cellulose and Other Natural Polymer Systems: Biogenesis, Structure, and Degradation;* Brown, R. M., Ed.; Plenum: New York, 1982; pp 243–271.
11. Basra, A. S.; Malik, C. P. *Int. Rev. Cytol.* **1984,** *89,* 65–113.
12. Meinert, M. C.; Delmer, D. P. *Plant Physiol.* **1977,** *59,* 1088–1094.
13. Delmer, D. P. *Annu. Rev. Plant Physiol.* **1987,** *38,* 259–292.
14. Timpa, J. D.; Triplett, B. A. *Planta* **1993,** *189,* 101–108.
15. Timpa, J. D. In *Cotton Fiber Cellulose: Structure, Function and Utilization;* National Cotton Council: Memphis, TN, 1993; pp 376–382.
16. Timpa, J. D.; Meredith, W. R. *Proceedings of the Beltwide Cotton Production Research Conferences;* National Cotton Council: Memphis, TN, 1993; Vol. 3, p 1556.
17. Timpa, J. D.; Zeronian, S. H. *Proceedings of the Beltwide Cotton Production Research Conferences;* National Cotton Council: Memphis, TN, 1993; Vol. 3, p 1493.
18. Wasserman, B. P.; Timpa, J. D. *Starch/Starke* **1991,** *43,* 389–392.
19. Politz, M.; Timpa, J. D.; Wasserman, B. R. *Cereal Chem.* **1994,** *71,* 532–536.
20. Politz, M.; Timpa, J. D.; White, A. R.; Wasserman, B. R. *Carbohydr. Polym.* **1994,** *24,* 91–99.
21. O'Donoghue, E. M.; Huber, D. J.; Timpa, J. D. *Planta* **1994,** *194,* 573–584.
22. Timpa, J. D. *Trends Polym. Sci.* **1993,** *1,* 105–110.

RECEIVED for review January 6, 1994. ACCEPTED revised manuscript April 26, 1994.

Size-Exclusion Chromatography of Dextrans and Pullulans with Light-Scattering, Viscosity, and Refractive Index Detection

William S. Bahary, Michael P. Hogan, Mahmood Jilani, and Michael P. Aronson

Unilever Research, 45 River Road, Edgewater, NJ 07020

The objective of this work was to characterize pullulans and dextrans in order to assess the performance of the size-exclusion chromatography (SEC) system with multi-angle laser light-scattering (MALLS), viscosity, and refractive index (RI) detectors using a buffered aqueous medium as the mobile phase. In 0.2 M $NaNO_3$, with standard pullulans that had low polydispersity, the weight-average molecular weights ($\langle M_w \rangle$) obtained by light scattering and by universal calibration agreed well with the expected values from the supplier with a mean deviation of about 5%. With standard dextrans that had higher polydispersities (1.6), the M_ws exhibited greater mean deviation, about 12%. Mark–Houwink exponents of 0.64 and 0.39 for pullulans and dextrans, respectively, obtained from plots were reasonable. The scaling law of radius of gyration versus $\langle M_w \rangle$ gave exponents of 0.48 and 0.38 for the polymers after adequate filtration. These were lower than the slopes calculated theoretically using the Ptitsyn–Eisner equation, but similar values have been reported elsewhere. Overall, triple detection SEC proved to be a powerful and time-saving technique for characterizing water-soluble polymers absolutely and confirming their values.

ALTHOUGH SIZE-EXCLUSION CHROMATOGRAPHY (SEC) with the conventional or universal calibration technique in combination with viscosity detection is well established in nonaqueous media (*1, 2*), its application in aqueous media is more complicated because of unusual difficulties

0065–2393/95/0247–0151$12.00/0
© 1995 American Chemical Society

associated with water-soluble polymers and biopolymers in aqueous systems (3, 4). These complications include hydrogen bonding, hydrophobic interactions, polar interactions, and ion–ion interactions that can make the separation by size nonideal. The first three can lead to adsorption or delay of sample elution times. These effects can be suppressed through the proper selection of column and mobile phase. Ion–ion interactions of the polymer with column support can lead to ion-exclusion and early elution relative to noninteracting polymers. Intraionic interactions of polyelectrolytes can lead to different polyion sizes and shapes with different elution times. These ionic effects can be minimized by the addition of salt to the mobile phase.

For samples for which no standards exist, low-angle laser light scattering (LALLS) is a possible solution, but has the drawback of excessive scattering by dust at low angles and it does not provide the macromolecular radius (5). Multi-angle laser light-scattering (MALLS) photometers are a potentially more powerful detection technique (6, 7). Therefore the objective of this work was to characterize pullulans and dextrans to assess the performance of the size-exclusion chromatograph with MALLS and viscosity and refractive index (RI) detectors. It was anticipated that with the combination of detectors, the performance of one detector could be effectively checked against that of the other. This type of checking becomes especially important in aqueous systems because in addition to the other issues, aqueous mobile phases are more difficult to clarify than nonaqueous ones.

Experimental Procedures

Materials. Standard pullulans having weight-average molecular weights ($\langle M_w \rangle$s) of 23.7, 48.0, 100, 186, and 380 kDa with a polydispersity (M_w/M_n) of about 1.1 were obtained from J. M. Science (Buffalo, NY). Standard dextrans having $\langle M_w \rangle$s of 40, 75, 170, 230, and 590 kDa with a polydispersity of about 1.6 were obtained from American Polymer Standards Corp. (Mentor, OH). The chemical structures of these polymers are displayed in Figure 1. The water used to prepare the mobile phase was 18 $M\Omega$-cm purified with the Barnstead Nanopure II apparatus. The other materials were reagent grade.

Equipment. The liquid chromatograph was a Waters 150C ALC/GPC operated at 30 °C equipped with 4 TSK PW columns (Chart I). A Wyatt Technology Corp. (Santa Barbara, CA) model Dawn F MALLS photometer was used with a Uniphase (San Jose, CA) argon-ion laser model 2011 using the 514.5-nm wavelength light. The viscosity detector was Viscotek model 100 (Porter, TX), and the differential refractometer of the 150C ALC/GPC was used as the RI detector. The Viscotek software employed was Unical versions 3.05 or 4.05 or as noted, and the Dawn Software Manual Version 1.01 was used.

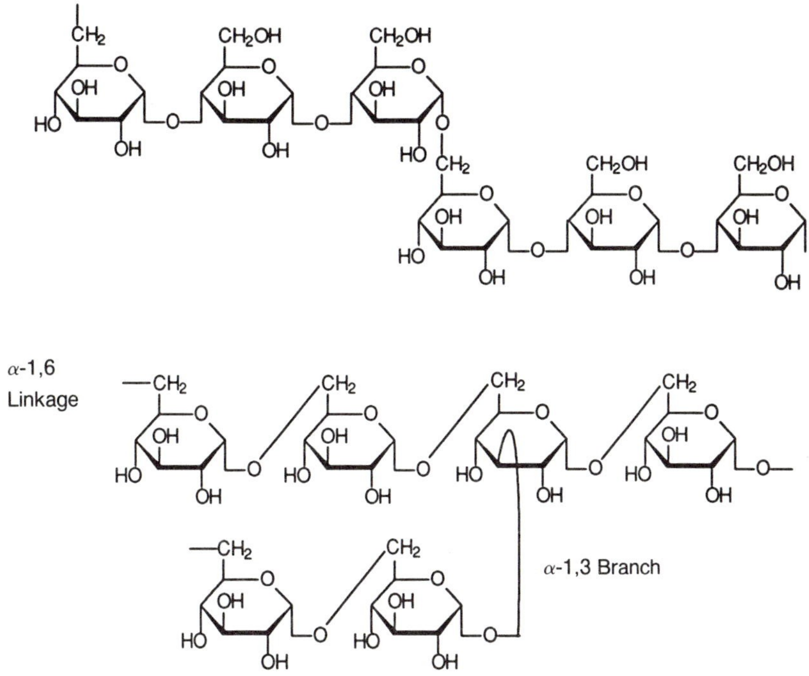

Figure 1. Chemical structure of pullulan containing maltotriose units and of dextran containing α-1,6 linked anhydroglucose units with α-1,3 branch points.

Procedures. The detectors were connected in series in the following sequence: SEC–MALLS–viscometer–RI–waste. This configuration was used on recommendation of Wyatt Corp. to maximize the RI signal of the very dilute solutions and to ascertain that both the LS and RI detectors observe the same mass in solution. The details of the experimental conditions are displayed in Chart I. The mobile phase employed was aqueous 0.2 M $NaNO_3$ that was clarified by filtration with 0.22-μm GS type filters (Millipore Corp., Bedford, MA), and the polymer solutions were filtered through 0.45-μm disposable nylon 66 syringe filters (Altec, Deerfield, IL). Light-scattering, intrinsic viscosity [η], and concentration chromatograms were obtained with 0.2 M $NaNO_3$ and analyzed as described later. The MALLS was calibrated with HPLC grade toluene. The normalization constants and the delay volumes were determined with a 23-kDa pullulan having a nominal radius of about 5 nm as a standard.

The RI constant, which relates the RI units to sample concentration, was determined by the dn/dc method of Wyatt Technology. The RI constant exhibited some variation at times, so that its value had to be adjusted in order to match the calculated mass with the injected mass. This variation was attributed to an intermittently faulty RI detector. The triple detection system had sufficient redundancy that accurate values of M_w and [η] could be obtained even when one component was not working well.

Chart I.   Experimental Conditions for SEC–LS–$[\eta]$

Liquid chromatograph	Waters 150C ALC/GPC
Detection	1. Wyatt Technology Dawn-F MALLS with argon-ion laser at 514.5 nm
	2. Viscotek differential viscometer
	3. Differential refractive index
Columns	TSK 7.5-mm × 30-cm columns
	1. G2500 PW (1×10^4 d, 100 Å)
	2. G3000 PW (1×10^5 d, 200 Å)
	3. G4000 PW (3×10^5 d, 500 Å)
	4. G5000 PW (1×10^6 d, 1000 Å)
Temperature	30 °C for GPC and viscometry
	RT for Dawn-F
Flow rate	1 mL/min
Sample concentration	1–3 mg/mL
Injection volume	0.1 mL
Mobile phase	0.2 M $NaNO_3$ in water
dn/dc at 514.5 nm	0.147 (±0.002) mL/g for dextran and pullulan in 0.2 M $NaNO_3$
Samples	Pullulan standards
	Dextran standards

Refractive Index Increment.   The RI differential, dn/dc, was determined with an Optilab model 903 (Wyatt Technology) at a wavelength of 514.5 nm. The value obtained for pullulans and dextrans as displayed in Chart I was 0.147 mL/g. In LDC/Milton Roy/Chromatix Technical Note LS 7, the value reported for 6.82×10^4 M_w dextran was 0.147 at 633 nm in water and 0.137 in 0.05 M KH_2PO_4 at pH 7 (8). Jackson et al. (5) used a value of 0.145 for pullulans and dextrans in 0.15 M NaCl. Vink and Dahlstrom (9) reported values of 0.148 and 0.145 for dextrans in water and in 0.5 M NaCl, respectively, at 546 nm, and 0.150 and 0.148 mL/g at 436-nm wavelength. Therefore the value obtained here is in line with other reported values.

Safety Precautions.   When using the argon-ion laser, several precautions should be followed.

1. Never look directly into the main laser beam.
2. Avoid contact with the high voltages of the laser head or power supply.
3. Use proper disposal procedures for broken or defective laser tubes.

The Uniphase laser manual should be consulted and followed carefully.

Results

Typical SEC, light-scattering, and viscosity data are presented first to indicate the type and quality of the results that were obtained.

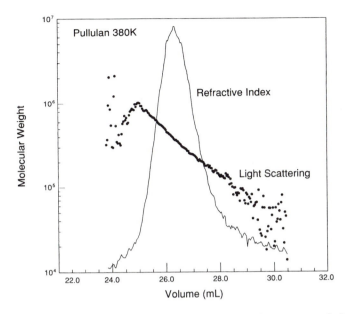

Figure 2. Light-scattering molecular weight and RI detector signal plotted against the retention volume.

In Figure 2, the light-scattering molecular weight and RI signal versus retention volume for the 380-kDa pullulan are displayed. The M_w decreases with the retention volume as expected, but much scatter occurs at the high and low ends and is attributed to the low concentrations at the ends. In Figure 3, the log intrinsic viscosity versus retention vol-

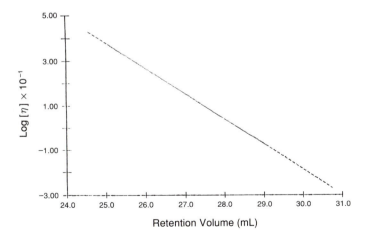

Figure 3. Log intrinsic viscosity smoothed by the Viscotek software versus the retention volume for pullulan 380 kD.

ume as smoothed by the Viscotek software to yield a linear fit is displayed; it suggests satisfactory fractionation.

In Figures 4 and 5, the M_ws from the light-scattering detector versus retention volume are displayed for three samples of pullulans and three dextrans, respectively. The molecular-weight calibration curves are smoother for the dextrans with broad molecular-weight distribution (MWD) than for the narrower pullulans, but both indicate adequate fractionation. Figure 6 displays the differential and cumulative molecular-weight distributions obtained by light scattering for the 380-kDa pullulan, and both indicate a reasonably narrow distribution with a short molecular-weight tail at the high end.

The slope of M_w versus retention volume as well as the polydispersity for narrow MWD samples was sensitive to the choice of delay volume as was reported by Shortt and Wyatt (10). In fact, Shortt recommended that one approach to determine the correct value of the delay volume is to adjust its value until an acceptable slope of the radius of gyration (R_g) versus M_w is obtained, because M_w is dependent on the concentration whereas the radius is not. Accordingly, the delay volume used was 0.312 mL.

Having established that the fractionation was satisfactory for the individual pullulan and dextran samples in 0.2 M NaNO$_3$, the light-scattering and viscosity results are viewed as a whole. The universal calibration plots for the pullulans and dextrans were superimposable,

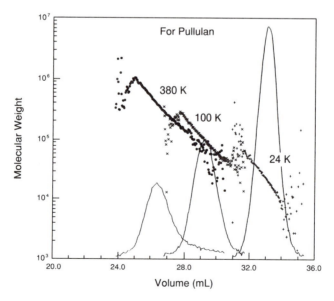

Figure 4. Log molecular weight versus retention volume for three pullulan samples obtained by MALLS.

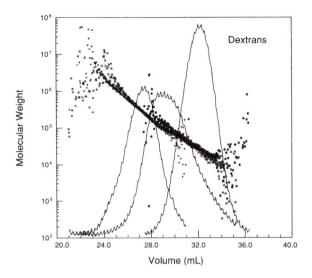

Figure 5.   Log molecular weight versus retention volume for the five dextran samples as obtained by MALLS.

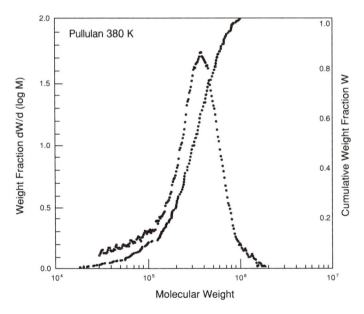

Figure 6.   Differential and cumulative molecular weight distributions for pullulan 380 kD obtained by SEC–MALLS.

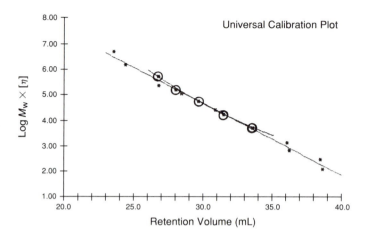

Figure 7. *Universal Calibration plots for pullulans (circles) and dextran (stars) obtained by SEC–Viscotek viscometer.*

as shown in Figure 7 so that $\langle M_w \rangle$s could be calculated from this plot. The narrow standards calibration method, which uses the peak M_w, was used for pullulans, whereas the broad standards calibration method, which uses M_w and M_n, was chosen for the broad dextran samples by the Viscotek software: hence the greater number of data points for the dextrans. The weight-average molecular weights of the pullulans obtained by the universal calibration method, as well as by light scattering, are displayed in Figure 8 and compared to M_w values provided by the supplier, which were obtained by equilibrium sedimentation. The agreement in the values obtained by all three techniques was remarkable, with a mean deviation of about 5% from the expected quantities. This result suggested that the system was operating satisfactorily.

The Mark–Houwink plot for the pullulans is displayed in Figure 9 and indicates a smooth relationship with little scatter. A slope of 0.64 was obtained from the best fit. Figure 10 displays a double logarithmic plot of the radius of gyration versus the molecular weight for pullulans, and this plot has a slope of 0.37. The theoretical values of R_g were calculated by using the Ptitsyn–Eisner equation (as follows) and are shown in the same figure (11):

$$[\eta]M = \phi(1 - 2.63\varepsilon + 2.86\varepsilon^2)(6^{1/2}R_g)^3$$

where $\phi = 2.86 \times 10^{23}$, $\varepsilon = (2a - 1)/3$, and a is the Mark–Houwink exponent.

Similar data for the standard dextrans, which had broader MWD than the pullulans, are presented in Figures 11–13. Figure 11 displays the weight-average molecular weights obtained from light scattering as well as the universal calibration plot and compares them to the expected values provided by the supplier. The M_ws exhibited a greater mean de-

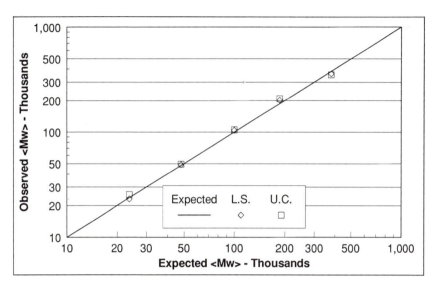

Figure 8. Comparison of pullulan molecular weights obtained by light scattering and universal calibration in relation to expected values from the supplier.

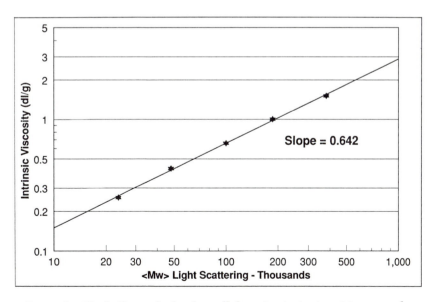

Figure 9. Mark–Houwink plot for pullulans. Intrinsic viscosities were obtained from the viscosity detector and $\langle M_w \rangle s$ from the light scattering one.

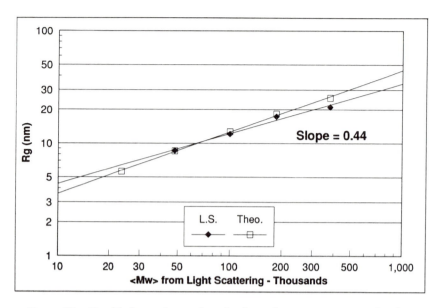

Figure 10.   Double logarithmic plot of radius of gyration versus molecular weight for pullulans. Theoretical slope obtained from Ptitsyn–Eisner equation (see text).

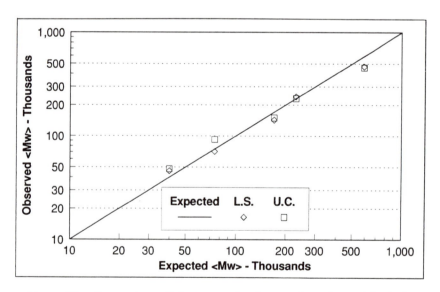

Figure 11.   Comparison of dextran molecular weights obtained by light scattering and universal calibration in relation to expected values from the supplier.

viation, about 12%, which is not unexpected, considering that the dextrans had higher polydispersities of about 1.6. Also, the molecular weights provided by the supplier may be questionable.

Figure 12 displays the Mark–Houwink plot for dextrans. The graph exhibits a slope of 0.39, which suggests extensive branching as reported elsewhere (*12*) and described later.

Figure 13 displays the log of the weight-average root-mean-square radius of gyration obtained from light scattering plotted against the log molecular weight for the dextran samples. The double logarithmic plots of R_g versus molecular weight for both the pullulans and dextrans exhibited excessive scatter below 10 nm, which is the lower limit of the light-scattering angular plot, $\lambda/20$. However, when the R_gs for the lower $\langle M_w \rangle$ samples were calculated by the Ptitsyn–Eisner equation, linear plots having theoretical slopes of 0.55 and 0.46 were obtained with all the data for pullulans and dextrans (Figures 10 and 13, respectively), as expected. Although the applicability of the Ptitsyn–Eisner equation to branched polymers may be open to question, the values obtained provided a reasonable guide for the low-molecular-weight samples that were beyond the lower limit of the light-scattering apparatus. Thus the light-scattering and viscosity detectors complemented each other.

Figure 14 displays double logarithmic plots of experimental R_g versus M_w for an extended range of M_ws of pullulans and dextrans from a second

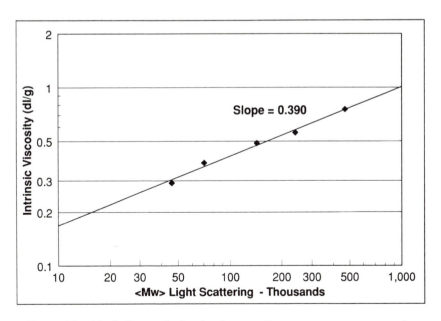

Figure 12.   Mark–Houwink plot for dextran. Intrinsic viscosities were obtained from the viscosity detector and $\langle M_w \rangle$s from the light scattering one.

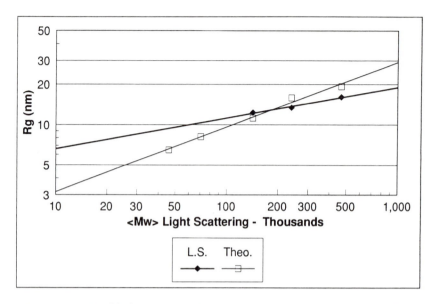

Figure 13.   Double logarithmic plot of radius of gyration versus molecular weight for Dextran. The theoretical values were obtained by the Ptitsyn–Eisner equation (see text).

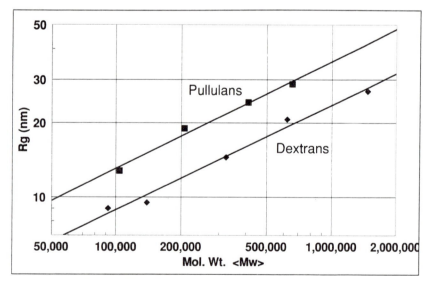

Figure 14.   Double logarithmic plot of radius of gyration versus molecular weight for pullulans and dextrans.

set of experiments in which the solutions were more carefully clarified. The slopes were found to be 0.48 and 0.38, respectively.

Discussion

Mark–Houwink Relations. The Mark–Houwink relationships obtained for pullulans and dextrans in 0.2 M $NaNO_3$ are as follows:

- for pullulan, $[\eta] = 4.22 \times 10^{-4} M^{0.64}$
- for dextran, $[\eta] = 4.85 \times 10^{-3} M^{0.39}$

For pullulans, Tsujisaka and Mitsuhashi report K and a values of 2.36 $\times 10^{-4}$ and 0.66 in water where the intrinsic viscosity is in deciliters per gram (*13, 14*). This result suggests that the saline is a better solvent than water for the pullulan and the agreement between the two values is good. The value for the exponent corresponds to that expected for a linear random coil polymer: about 0.5–0.8 depending on solvent power (*13*).

For linear fractions of dextrans in water, Senti, et al. (*12*) reported values of 97.8×10^{-3} and 0.50 for the constants, and for branched dextrans the value of the exponent was 0.20. Cerney, et al. (*16*) reported values of 10.3×10^{-3} and 0.25 for branched dextrans in methanol–water. The suppliers' data in 0.05 M Na_2SO_4 yields values of 9 $\times 10^{-4}$ and 0.50 at 30 °C for a broad range of molecular weights (*17*). The results of this work are in line with those reported for dextrans, and the exponent corresponds to that for a highly branched coil about 0.0–0.5 (*18*).

For an individual pullulan sample, the exponent obtained with the revised Viscotek software Unical 3.05 or 4.05 was 0.66 and agrees well with the value for the whole polymer. According to Viscotek, the later versions of their software forces the values to agree by adjusting the concentration term in the calculations for narrow MWD samples because the range of intrinsic viscosities is narrow and subject to large errors. For this reason, the older software version 2.70 was modified because it gave erroneously low values for the exponent with narrow MWD samples like pullulan. For the broad MWD dextran, the exponent for an individual sample was 0.36 with all the software versions and agrees well with that for the whole polymer samples. These results are summarized in Table I.

Radius versus M_w Relation. The conformational coefficients a_r from the double logarithmic plots of R_g versus $\langle M_w \rangle$ for pullulans and dextrans were 0.44 and 0.30, respectively, and were low (Figures 10 and 13). In the second experiment in which the mobile phase and so-

Table I.   Comparison of Scaling Parameters for Pullulans and Dextrans

| | $[\eta] = KM^{a_\eta}$ | | $R_g = K'M^{a_R}$ | |
| | a_η | | a_R | |
Sample	Expected	Observed	Expected	Observed
Pullulans				Expt. 1 Expt. 2
Whole polymers	0.50–0.80	0.64	0.50–0.60	0.44 0.48
P-380K slices		0.67		0.30 0.45
Dextrans				
Whole polymers	0.0–0.50	0.39	0.33–0.50	0.23 0.38
DX-590K slices		0.36		0.32 0.32

lutions were clarified better, the correlation coefficients were 0.48 and 0.38, which are more reasonable because the minimum value is 0.33. Jackson et al. (5) reported values 0.45 and 0.36 in 0.15 M sodium nitrate which are comparable to the values reported here but lower than the theoretical ones calculated from the Ptitsyn–Eisner equation, 0.55 and 0.47. Fishman et al. (19) reported values of 0.59 and 0.43 from light scattering for whole polymers in water where the z-average root-mean-square radius of gyration R_z was used.

By way of explanation, the low a_r values observed reflect the difficulty in clarifying aqueous solutions for light scattering. For the low M_w samples, the dust particles can mask the angular dependence of scattered light by the small macromolecules. For the higher M_w samples, the lower angle detectors were routinely omitted, so that any curvature in the angular dependence would have been missed. Extreme care in using dust-free water, as well as in clarifying the mobile phase and sample solutions, is recommended. Furthermore, recycling the mobile phase may be helpful in obtaining more reliable and accurate results.

For individual samples of pullulans and dextrans, the conformational coefficients of the R_g versus M_w plots were 0.45 and 0.32, respectively, and are in reasonable agreement with the values for the whole sample as displayed in Table I. These are sensitive to the value of the delay volume as described earlier. Minimizing the delay volume and using bovine serum albumin, which is monodisperse, to determine the delay volume may be helpful.

It appears that for low values of R_g (10–20 nm), the values of R_g that were calculated from the Ptitsyn–Eisner equation were more reliable than those obtained from light scattering in an aqueous medium. Accordingly, for accurate work over a broad range of molecular weights and sizes, a triple detection system is advantageous and recommended.

Summary

In summary, a SEC instrument was set up with light scattering, viscosity, and RI detectors. The $\langle M_w \rangle$s by L.S. and universal calibration agreed very well with expected values for the low polydispersity pullulans and reasonably well for the broad MWD dextrans. The R_g from light scattering agreed well with the theoretical R_g for values above 10 nm, and scaling law parameters in terms of viscosities or radii for whole polymers were in reasonable agreement with expected values when the aqueous solutions were adequately clarified. In conclusion, triple detection SEC proved to be a powerful and time saving technique for characterizing water soluble polymers absolutely, and confirming the values as well.

Acknowledgement

The authors are grateful to Unilever Research for permission to publish this work.

References

1. Yau, W. W.; Kirkland, J.; Bly, D. D. *Modern Size Exclusion Liquid Chromatography*; Wiley: New York, 1979; pp 291–294 and 335–338.
2. Grubisic, Z.; Remp, P.; Benoit, H. *J. Polym. Sci., Part B* **1967**, *5*, 753.
3. Bahary, W. S.; Jilani, M. *J. Appl. Polym. Sci.* **1993**, *48*, 1531.
4. Nagy, D.; Terwilliger, D.; Lawrey, B.; Tiedge, W. F. In *International GPC Symposium '89 Proceedings*, Waters, Division of Millipore Corp.: Newton, MA, 1989; pp 637–662.
5. Jackson, C.; Nilsson, L.; Wyatt, P. *J. Appl. Polym. Sci.: Appl. Polym. Swamp.* **1989**, *43*, 99.
6. Wyatt, P. J.; Jackson, C.; Wyatt, G. K. *Am. Lab.* **1988**, May–June.
7. Williams, D. L.; Pretus, H. A.; McNamee, R.; Jones, E. In *International GPC Symposium '91 Proceedings*; Waters, Division of Millipore Corp.: Newton, MA, 1991.
8. LDC/Milton Roy/Chromatix Technical Note LS 7, Division of Thermo Separation Products, Rivera Beach, FL, 1979.
9. (a) Vink, H.; Dahlstrom, G. *Makromol. Chem.* **1967**, *109*, 249 (b) *See* also *Polymer Handbook*, 3rd ed.; Brandrup, J.; Immergut, E., Eds.; Wiley: New York, 1989; pp vii, 469.
10. Shortt, D.; Wyatt, P. Presented at Waters International GPC Symposium '94, Orlando, FL, June 5–8, 1994.
11. Yau, W. W.; Kirkland, J.; Bly, D. D. *Modern Size Exclusion Liquid Chromatography*; Wiley: New York, 1979; p 35.
12. Senti, F. R.; Hellman, N. N.; Ludwig, N. H.; Babcock, G. E.; Tobin, R.; Glass, C. A.; Lamberts, B. L. *J. Polym. Sci.* **1955**, *27*, 527–546.
13. Tsujisaka, Y.; Mitsuhashi, M. In *Industrial Gums*; Whistler, R.; BeMiller, J., Eds.; Academic: Orlando, FL, 1993; p 450.
14. Kawahara, K.; Ohta, K.; Miyamoto, H.; Nakamura, S. *Carbohyd. Polym.* **1984**, *4*, 335.
15. Tanford, C. *Physical Chemistry of Macromolecules*; Wiley: New York, 1967; pp 407–408.

16. Cerney, L. C.; McTiernan, J.; Stasiw, D. *J. Polym. Sci.* **1973**, *42*, 1455–1465.
17. *Polymer Standards Catalog;* American Polymer Standards Corp.: Mentor, OH, February 8, 1993.
18. Zimm, B.; Kilb, R. *J. Polym. Sci.* **1959**, *37*, 19.
19. Fishman, M.; Damert, W.; Phillips, J.; Bradford, R. *Carbohyd. Res.* **1987**, *160*, 215–225.

RECEIVED for review January 6, 1994. ACCEPTED revised manuscript July 22, 1994.

13

Star-Branched Polymers in Multidetection Gel Permeation Chromatography

James Lesec,[1] Michèle Millequant,[1] Maryse Patin,[2] and Philippe Teyssie[2]

[1]Université Paris VI—Centre National de la Recherche Scientifique, Unité de Recherche Associée 278, Ecole Supérieure de Physique et de Chimie Industrielles de la Ville de Paris, lo rue Vauquelin, 75231 Paris cedex 05, France
[2]Institut de Chimie, Centre d'Etude et de Recherche sur les Macromolécules—Université de Liège, Sart-Tilman B6, 4000 Liège, Belgium

Multidetection gel permeation chromatography (GPC) is a very powerful tool for the study of complex polymers. This chapter deals with the characterization of star-branched model copolymers using refractive index and viscometric detection with an on-line light-scattering detector. The polymer branches are composed of methyl methacrylate–tert-butyl acrylate diblock copolymers of poly(methyl methacrylate) and poly(tert-butyl acrylate) with a very well-controlled chemical composition and structure (very low polydispersity). When these branches are chemically coupled, they produce star-branched copolymers with a various number of branches. These copolymers were synthesized at Liege University in Belgium and our problem was to characterize these macromolecules. In GPC, viscometric detection allows the determination of molecular weights and long-chain branching distribution when used with the universal calibration curve. The low-angle laser light-scattering (LALLS) detector provides absolute molecular weights. It has been found that excellent agreement can be obtained from viscometry with universal calibration and from the LALLS detector when determining molecular weights. The results demonstrate an excellent performance for the universal calibration, even for highly branched polymers with a very particular viscometric behavior. Number-average molecular weight ($\overline{M}_n$) values from stars and branches were used to determine the number of branches of the star-branched copolymers. The values obtained with the viscometric detection and the LALLS detector were compared and found to be

0065–2395/95/0247–0167$12.00/0

similar. Several different copolymers with various numbers of branches were studied.

GEL PERMEATION CHROMATOGRAPHY (GPC) is a very powerful tool for the characterization of polymers. For the study of complex polymers with a very complicated structure, the use of multidetection techniques greatly increase the power of characterization. Generally, mass detectors, like viscometric detectors and light-scattering detectors, are used to get more information about the polymer. A Waters 150 CV (Waters Corporation, Milford, MA) (differential refractive index (DRI) and single capillary viscometric detection) with an on-line light-scattering detector (LALLS-Chromatix CMX 100, Thermo Separation Products, Riviera Beach, FL) were used for the characterization of star-branched model copolymers. The coupling of the DRI with the viscometric detection allows the determination of branching distribution and the calculation of average molecular weights using a universal calibration curve. Also, the coupling of the RI with light-scattering detection provides the average molecular weights.

We wanted to determine the number of branches of star-branched model copolymers, with the number average Mn as the key parameter. Light-scattering detection is not the most appropriate method to determine the number-average molecular weight $\overline{M}_n$ but provides the weight-average molecular weight $\overline{M}_w$ with accuracy. The viscometric detection provides results, assuming that the universal calibration is perfectly observed. In this work, we used light-scattering detection to determine $\overline{M}_w$ and to compare these values to those obtained by viscometric detection. The comparison of $\overline{M}_w$ validates the use of the universal calibration curve and, therefore, the use of the number averages $\overline{M}_n$ coming from the viscometric detection to characterize the star-branched copolymers.

Materials and Methods

The star-branched polymers are copolymers of poly(methyl methacrylate) (PMMA) and poly(*tert*-butyl acrylate) (PtBuA) (Figure 1). The polymer branches are composed of PMMA–PtBuA diblock copolymers with well-controlled chemical composition and structure. They are synthesized by the usual sequential anionic addition polymerization using a method previously described (1) that provides diblock copolymers with a very low polydispersity. The star-branched copolymers are obtained by coupling the living linear diblock copolymers with a coupling agent, like ethylene glycol dimethacrylate or 1,6-hexanediol diacrylate (Figure 1). Several well-defined star block copolymers with different arm lengths and different chemical compositions were prepared by the usual sequential anionic addition polymerization. They are described in Table I.

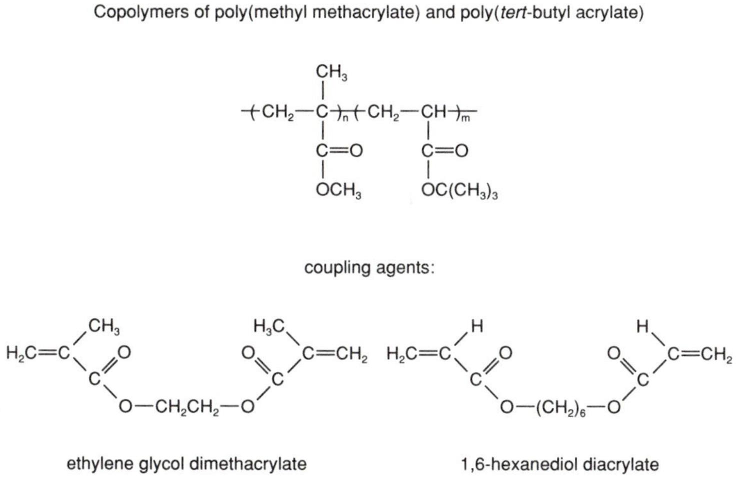

Copolymers of poly(methyl methacrylate) and poly(*tert*-butyl acrylate)

$$\begin{array}{c} CH_3 \\ | \\ -\!\!(CH_2\!\!-\!\!C\!\!)_n\!\!(CH_2\!\!-\!\!CH\!)_m \\ | \qquad\quad | \\ C\!\!=\!\!O \qquad C\!\!=\!\!O \\ | \qquad\quad | \\ OCH_3 \qquad OC(CH_3)_3 \end{array}$$

coupling agents:

ethylene glycol dimethacrylate   1,6-hexanediol diacrylate

Figure 1.   *Chemical structure of copolymers and coupling agents.*

For every sample, an aliquot of the reaction medium was drawn before the coupling reaction to obtain the linear copolymer corresponding to the arm of the star-branched copolymer. The use of both compounds facilitates the interpretation of the GPC experiments.

As the coupling reaction is not fully efficient, the star-branched copolymers contain a nonnegligible amount of uncoupled linear copolymer. To get the star-branched copolymers as pure as possible, a purification step where linear chains were extracted by fractional precipitation using different solvent or nonsolvent methods was necessary (*1*).

A Waters GPC 150 CV, equipped with the DRI prototype #4 and a single capillary viscometer, was used for this study. A low-angle laser light-scattering (LALLS) detector (Chromatix CMX 100) was inserted between the column set and the GPC 150 CV detectors. Tetrahydrofuran (THF) was used at 40 °C and at a flow rate of 1 mL/mn. THF was filtered on a Millipore membrane-type FH and stabilized by Ionol at a concentration of 0.04%. The columns used were a set of Waters Ultrastyragel (103–106 Å). The narrow standards used for calibration were a set of polystyrene standards

Table I.   Composition of Samples

Samples	$\overline{M}_n$ PMMA	$\overline{M}_n$ PtBuA	% PMMA in Branch
PM4	4150	18,250	22.5
PM1	6390	21,130	23.2
PMA	8120	9980	44.7
PM25	20,200	20,900	49.2
PM2	16,730	14,460	53.6
PM3	16,530	7400	68.2

SOURCE: Data are taken from reference 1.

(TSK) from Toyo Soda (Toyo Soda, Japan) in a molecular weight range of 3000–3,000,000.

A differential refractometer DRI prototype (#4) was used to avoid erroneous results that a standard DRI detector may cause in viscosity calculations because of very small flow fluctuations, so-called "Lesec effect," when the polymer flows across the detectors (2–4). This very small fluctuation is induced by the specific viscosity of the polymer solution and is enough to produce a significant apparent shift downstream of the viscometer peak. This shift leads to a small rotation of the viscosity–molecular weight relationship (2–4) and a decrease of the Mark–Houwink exponent for linear polymers or a distortion of the viscosity law for branched polymers. This has been previously described (2–4). DRI prototype (#4) has been especially designed to reduce this effect. A special geometry is used to reduce the pressure drop in the detector area but also to reduce the void volume. This design is used in the new Waters GPC 150 CVplus, and the details on the DRI design will be published later (5).

"Multidetector GPC software", a PC-DOS package written by Lesec (6), was used for triple detection GPC. For data acquisition, the personal computer is connected to the 150 CV and the CMX 100 through a CEC IEEE board (Capital Equipment Corporation, Burlington, MA) and a 199 scanner multimeter (Keithley, Cleveland, OH). Molecular weights are calculated using both the LALLS detection and the viscometric detection with a combination of the classical molecular weight calibration curve and the viscosity law of the standards (6).

Preparation of the GPC system, especially the performance of the viscometric detection and the performance of the refractometer DRI prototype (#4), was performed by the comparison of the viscosity law obtained using the TSK narrow standards represented in Figure 2 and the one obtained with a linear broad polymer, like the polystyrene Dow 1683 (Dow Chemical,

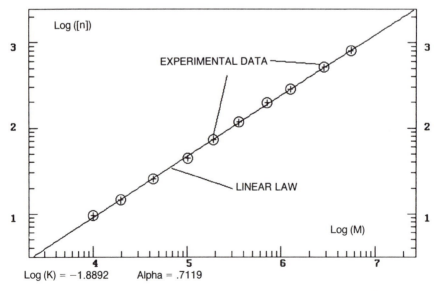

Figure 2. Viscosity law of TSK narrow polystyrene standards.

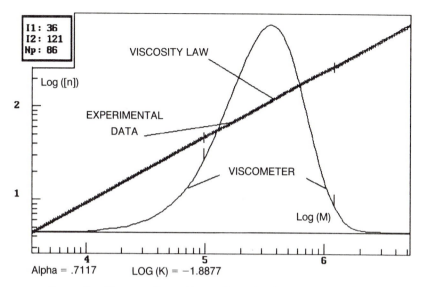

I1 : 36
I2 : 121
Np : 86

VISCOSITY LAW

Log ([n])

2

EXPERIMENTAL
DATA

VISCOMETER

1

Log (M)

4
Alpha = .7117

5
LOG (K) = −1.8877

6

Figure 3. Viscosity law of linear broad polystyrene Dow 1683.

Midland, MI) represented in Figure 3. Very similar laws are obtained, confirming the ability of this new GPC instrument to provide the right viscosity information.

The problem of determining the average number of branches is theoretically simple, it is the ratio of the $\overline{M}_n$ of the star polymer to that of the uncoupled linear branches. In fact, the branched copolymers, even after a heavy purification, contain a residual amount of unused linear branches. Because these two materials have different molecular weights, they elute separately, and it is possible to analyze the star peak alone. The big issue is that this "impurity" (uncoupled linear branches) is weighed when making the solution and, accordingly, the real concentration of the star copolymer is not known exactly; this information is very necessary when using GPC with mass detectors.

A method to determine the sample concentration independent of knowing the exact concentration can be used and consists of calibrating the DRI response R using a polystyrene sample with a well-known RI increment dn/dc (Dow 1683) to use the DRI relationship, where K is the calibration constant and C is the sample concentration:

$$R = K * dn/dc * C$$

Using the K constant determined by using the broad polystyrene DOW 1683, it is then possible to measure the dn/dc of each branch because they are very pure and their concentrations are perfectly known. Finally, assuming the dn/dc of the star copolymer is extremely close to the one of the branches (there are only very few amounts of coupling agent), it is just necessary to set the software in the "concentration correction" position to determine the correct slice concentration using the DRI signal, K, and $dn/$

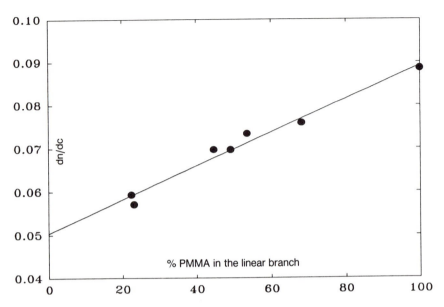

Figure 4. dn/dc of arm copolymers as a function of PMMA content.

dc. This corrected slice concentration is then used in the light-scattering calculations to determine the molecular weights and in the viscosity calculations to get the right slice intrinsic viscosity and also molecular weights. The dn/dc of the different branch copolymers are represented in Figure 4 as a function of the PMMA content.

Results and Discussion

Figure 5 represents the chromatograms of the linear arm copolymer corresponding to sample PM1. The three peaks nearly overlay completely, which indicates a very low polydispersity measured as 1.06. Figure 6 represents the chromatograms of the star-branched sample PM1. The LALLS peak is normally shifted toward the high molecular weight side, but the viscometer peak is very similar to the DRI one, which is very unusual. In a classical representation, represented in Figure 7 for the linear broad polystyrene Dow 1683, the viscometric response is very close to the light-scattering response, the first one being proportional to $C*[\eta]$ (i.e., $C*M^{0.7}$) and the second one being proportional to $C*M$ (C is the concentration, $[\eta]$ the intrinsic viscosity, and M the molecular weight). This particular behavior is due to the high long-chain branching that tremendously reduces viscosity. Intrinsic viscosity variations of PM1 star-branched copolymer versus elution volume are represented in Figure 8. A strong distortion occurs with regard to a linear macromolecule, due to the particular long-chain branching of this

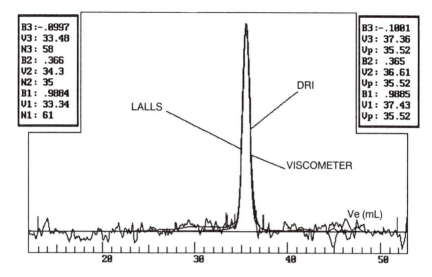

Figure 5.   Chromatograms of the arm copolymer corresponding to PM1.

sample. Also, intrinsic viscosity variations versus Log M are plotted in Figure 9, where the experimental viscosity law (viscosity–molecular weight relationship) is compared with the one of the branch (linear) to determine the branching distribution g_i (g_i is the ratio $[\eta]_{bri}:[\eta]_{lini}$, intrinsic viscosity of the branched and linear polymers, respectively, at the same molecular weight). These plots confirm the very branched behavior of

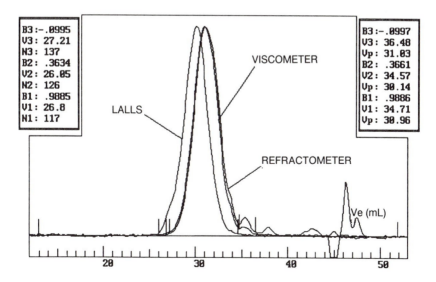

Figure 6.   Chromatograms of the star-branched copolymer PM1.

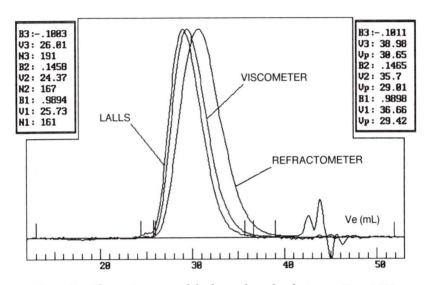

Figure 7.   Chromatograms of the linear broad polystyrene Dow 1683.

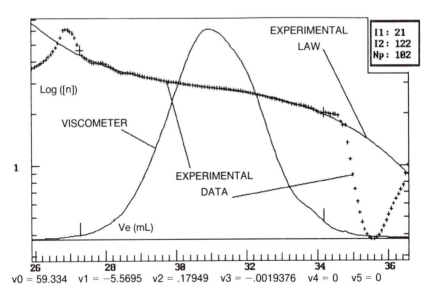

Figure 8.   Viscosity variations of the star-branched polymer PM1. PM1 viscosity peak and Log $[\eta]_i$ versus elution volume.

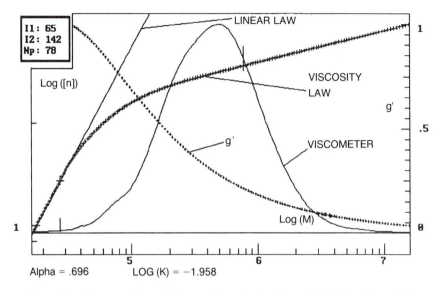

Figure 9. Viscosity law of the star-branched polymer PM1. Log [η]ᵢ and g'ᵢ branching distribution versus molecular weight.

these molecules because g'_i strongly decreases to around zero for high molecular weights. As a comparison, we have plotted in Figure 10 the same branching representation for a very well-known long-chain branched polymer, the low density polyethylene NBS 1476 (National Institute of Standards and Technology, Washington, DC). The comparison shows the difference in long-chain branching between the two samples, the PM1 sample being obviously much more branched than the classical NBS 1476.

Table II contains the results of the analysis of star-branched copolymers ($\overline{M}_w$, $\overline{M}_n$, and polydispersity d) by both the viscometric coupling (V) and the light-scattering coupling (L). The numerical values are in agreement, especially for the $\overline{M}_w$ values. These results confirm that the universal calibration is perfectly valid for branched molecules, even for a high degree of long-chain branching.

Although the light-scattering coupling is not the most appropriate method to measure the number average molecular weight $\overline{M}w_n$, because of the lack of information on the scattered light in the low molecular weight side, the agreement looks reasonable for both $\overline{M}_w$ and $\overline{M}_n$ values from GPC–viscometry and GPC–LALLS. These results confirm that the universal calibration works well for these kinds of macromolecules.

Table III contains the numerical results in Mn for the six different samples, for arms (lin) and stars (br). The values obtained by GPC–viscometry (V) and by GPC–LALLS (L) are in agreement. They ap-

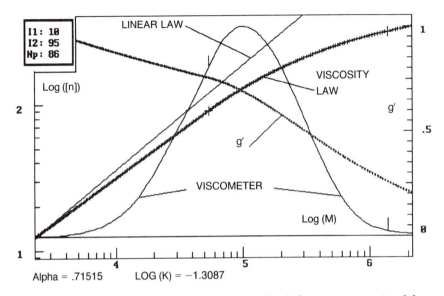

Figure 10.   Viscosity law of the branched polyethylene NBS 1476. Log [η]ᵢ and g'ᵢ branching distribution versus molecular weight.

proximately lead to the same number of branches per star [Br–Lin(V) and Br–Lin(L)].

Considering the results from Tables II and III, it is difficult to determine which parameter, Br–Lin(V) or Br–Lin(L), seems the most appropriate to characterize the average number of branches of our star-branched copolymers, which roughly varies from 3 to 12.

The intrinsic viscosity $[\eta]$ of both linear and branched macromolecules and the branching parameter $<g'>$ values of the star-branched copolymers were also calculated by the GPC software and are reported in Table IV. The ratio $[\eta]_{br}:[\eta]_{lin}$, between the intrinsic viscosity of the

Table II.   Characterization of Star-Branched Copolymers in $\overline{M}_n$ and $\overline{M}_w$

Samples	$\overline{M}_{w_{lin}}(V)$	$\overline{M}_{n_{lin}}(V)$	$d(V)$	$d(L)$	$\overline{M}_{w_{br}}(L)$	$\overline{M}_{n_{br}}(L)$
PM4	209,400	125,400	1.67	1.61	177,000	109,900
PM1	472,600	261,800	2.18	2.31	548,200	237,300
PMA	753,000	208,600	3.61	4.15	782,300	188,500
PM25	295,400	202,300	1.46	1.54	280,000	181,800
PM2	132,100	82,020	1.61	1.66	140,900	84,860
PM3	171,200	114,900	1.49	1.36	143,300	105,400

Table III. Number of Branches Per Star by GPC–Viscometry
and GPC–LALLS

Samples	$\overline{M}_{n_{lin}}(V)$	$\overline{M}_{n_{br}}(V)$	Br–$Lin(V)$	Br–$Lin(L)$	$\overline{M}_{n_{lin}}(L)$	$\overline{M}_{n_{br}}(L)$
PM4	20,530	125,400	6.1	6.9	16,030	109,900
PM1	26,740	261,800	9.8	10.3	22,930	237,300
PMA	18,080	208,600	11.5	10.9	17,250	188,500
PM25	37,560	202,300	5.4	5.4	33,850	181,800
PM2	24,560	82,020	3.3	4.0	21,370	84,860
PM3	21,190	114,900	5.4	5.6	18,820	105,400

star and the one of the linear molecule used to synthesize the star, is found to be approximately constant and equal to 2. This interesting result has already been observed (7) for other star-branched polymers. It indicates that the size of the star-branched copolymers is a particular function of the number of branches.

Figure 11 represents the variations of $<g'>$ as a function of the number of branches. The $<g'>$ value strongly decreases when the number of branches in the stars increases. Again, a very good agreement is observed between GPC–viscometry and GPC–LALLS values.

Conclusion

The characterization of star-branched polymers has been performed using triple detection because it was not obvious, in the beginning of this study, that universal calibration could be applied to star-branched polymers. In fact, the GPC software used handles experimental data as a double dual detection, GPC–viscometry and GPC–LALLS. Experimentally, it has been found that excellent agreement between results from these two sets of data can be obtained. GPC–viscometry uses a universal calibration curve and GPC–LALLS is free of any molecular weight calibration curve. Therefore, the universal calibration works well with very long chain branched polymers, even with a very particular

Table IV. Intrinsic Viscosity and Branching Parameter $\langle g' \rangle$
by GPC–Viscometry

Samples	$[\eta]_{lin}$	η_{br}	$[\eta]_{br}:[\eta]_{lin}$	$\langle g' \rangle br$
PM4	11.87	21.62	1.82	0.404
PM1	13.05	27.63	2.12	0.283
PMA	10.47	20.12	1.92	0.168
PM25	18.07	33.82	1.87	0.496
PM2	13.04	23.50	1.80	0.630
PM3	11.83	21.69	1.83	0.458

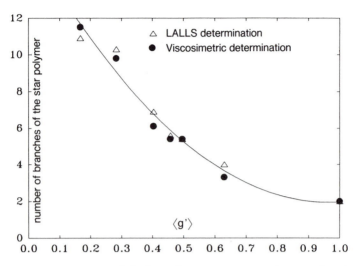

Figure 11.   Variations of the branching parameter <g'> as a function of the number of branches in the star-branched copolymer PM1.

viscometric behavior. The calculation of the number of branches was performed by making the ratio of the number average molecular weight of the star polymer to that of the uncoupled linear branches. The use of Mn from GPC–viscometry and Mn from GPC–LALLS leads approximately to the same number of branches in the star-branched copolymers. The ratio $[\eta]_{br}:[\eta]_{lin}$, between the intrinsic viscosity of the star and the one of the linear molecule used to synthesize the star, is found to be approximately constant and equal to 2, which has already been observed for other star-branched polymers.

References

1. Fayt, R.; Jacobs, C.; Jerome, R.; Ouhadi, T.; Teyssie, P.; Varshney, S. K. *Macromolecules* **1987**, *20*, 1442; Varshney, S. K.; Hauteheer, J. P.; Fayt, R.; Jerome, R.; Teyssie, P. *Macromolecules,* **1990**, *23*, 2618.
2. Lesec, J. Presented at the First International Symposium on GPC/Viscometry, Houston, TX, 1991.
3. Huard, T.; Dark, W. A.; Ekmanis, J. L.; Havard, T. J.; Nielson, R.; Lesec, J. Presented at the First International Symposium on GPC/Viscometry, Houston, TX, 1991; and *Proceedings of the Waters International GPC Symposium 91;* Waters: Milford, MA, 1991; pp 285, 294; Lesec, J.; Millequant, M.; Havard, T. *Am. Chem. Soc., Polym. Mater. Sci. Eng.* **1991**, *65*, 138; *Gel Permeation Chromatography: Characterization by SEC and FFF;* Provder, T., Ed.; ACS Symposium Series 521; American Chemical Society: Washington, DC, 1991.

4. Lesec, J. *Proceedings of the Waters International GPC Symposium 91;* Waters: Milford, MA, 1991.
5. Nielson, R.; Lesec J. unpublished results.
6. Lesec, J.; Volet, G. *J. Liq. Chromatogr.* **1990,** *13,* 831; *J. Appl. Polym. Sci. Appl. Polym. Symp.* **1990,** *45,* 177.
7. Zilliox, J. G.; Rempp, P.; Parod, J. *J. Polym. Sci. C* **1968,** *22,* 145.

RECEIVED for review January 6, 1994. ACCEPTED revised manuscript June 23, 1994.

ANALYSIS OF COMPOSITIONAL HETEROGENEITY IN COPOLYMERS AND BLENDS

Determination of Molecular Weight and Composition in Copolymers Using Thermal Field-Flow Fractionation Combined with Viscometry

Martin E. Schimpf

Department of Chemistry, Boise State University, Boise, ID 83725

A method is presented for obtaining the molecular weight and composition of copolymers using thermal field-flow fractionation (ThFFF) and viscometry. The method requires the thermal diffusion coefficients for the homopolymer constituents, which are constant in a given solvent and can be measured independently using ThFFF (and have already been determined for many polymer-solvent systems). Equations are derived that express the average molecular weight and composition as a function of ThFFF retention and intrinsic viscosity in two separate solvents. The method is demonstrated in two pairs of solvents using a statistical copolymer of styrene and isoprene, but it is also expected to apply to linear-block copolymers whose secondary structure is randomized by nonselective solvents (i.e., solvents that are equally good for both copolymer components). An analysis of errors indicates that the accuracy of the method primarily depends on the precision with which ThFFF retention is measured, as well as reliable values for the thermal diffusion coefficients.

THERMAL FIELD-FLOW FRACTIONATION (ThFFF) separates polymers according to their molecular weight and chemical composition. The molecular weight dependence is well understood and is routinely used to characterize molecular weight distributions (*1–4*). However, the dependence of retention on composition is tied to differences in the thermal diffusion of polymers, which is poorly understood. As a result, the compositional selectivity of ThFFF has not realized its full potential. How-

0065–2395/95/0247–0183$12.00/0

ever, we demonstrated the ability to predict thermal diffusion (hence retention) in certain copolymers (5), allowing us to use ThFFF to obtain compositional information in such copolymers.

Background

It was recognized early in the development of ThFFF that different molecular weight components are separated because of the dependence of ordinary (Fickian) diffusion on hydrodynamic radius, whereas thermal diffusion is responsible for variations in retention with polymer composition (6). The ability to separate polymers based only on compositional differences was first demonstrated by Gunderson and Giddings (7) when poly(methyl methacrylate) and polystyrene polymers that were similar in size, and therefore coeluted in size-exclusion chromatography, were separated in a ThFFF channel.

Although the relationship between ordinary diffusion and molecular weight is well understood, we are still struggling to understand how composition affects thermal diffusion, and therefore ThFFF retention. Although we can relate retention to the phenomenological coefficient for thermal diffusion (D_T), D_T has not been successfully related to physicochemical parameters of the polymer and solvent. As a result, retention cannot be related to polymer composition unless D_T is first determined empirically.

To improve our ability to predict ThFFF retention from compositional information, or vise versa, we are studying thermal diffusion in a systematic manner using ThFFF itself, which is capable of yielding precise values of D_T. These studies have shown that in homopolymers, thermal diffusion is independent of molecular weight and branching structure (8) but significantly affected by the composition of both polymer and solvent (9). The independence of thermal diffusion and molecular weight is fortuitous, as it reduces ThFFF calibrations to a single measurement, provided the relationship between either diffusion or viscosity and molecular weight is known (10). Unfortunately, the compositional dependence of retention in homopolymers cannot yet be predicted.

The thermal diffusion of several copolymers was examined in a variety of solvents (5, 11). For statistical (random) copolymers, D_T is a weighted average of the D_T values of homopolymers made of the constituent monomers, as illustrated in Figure 1. For example, in a given solvent, the D_T value of a statistical copolymer containing 50 mole percent styrene and 50 mole percent isoprene is a simple average of the D_T values of polystyrene and polyisoprene homopolymers. For copolymers with a greater styrene content, D_T is weighted more toward that of polystyrene homopolymer, where the weighting factor is the mole fraction of styrene in the copolymer.

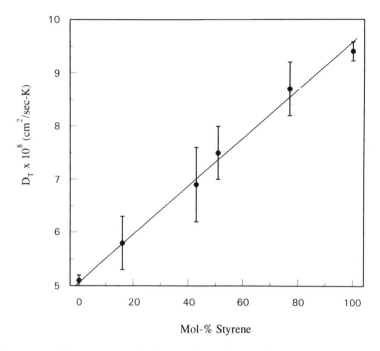

Figure 1. Illustration of the linear dependence of D_T on composition in copolymers of styrene and isoprene. In addition to the two homopolymer end-points, one statistical copolymer and three linear-block copolymers are represented.

The situation is more complex in block copolymers. Thermal diffusion, and therefore retention, can still be predicted in block copolymers, provided the monomeric units are not radially segregated in the solvated coil. Segregation is critical because monomeric units located in the outer free-draining region of the solvated coil appear to dominate thermal diffusion. In statistical copolymers, monomer segregation is not possible. In block copolymers, on the other hand, segregation can arise from bonding constraints or solvent effects (*11*). However, as long as a nonselective solvent is used (a solvent that is equally good for all copolymer blocks), the monomeric units of linear-block copolymers will not segregate appreciably and thermal diffusion follows the same behavior as it does in statistical copolymers. As a result, retention becomes predictable in linear-block copolymers when a nonselective solvent is used. Finally, in highly branched block copolymers, randomization of the monomeric units is physically restricted, so that retention cannot be predicted even in nonselective solvents (5).

In summary, the methods described here can be used to determine both composition and molecular weight in copolymers where D_T is a linear function of composition. This is the case with statistical copolymers and, if two nonselective solvents are available, for linear-block copolymers as well.

Finally, it should be noted that D_T varies linearly with the temperature of the cold wall T_c (12). For example, D_T values of polystyrene in ethylbenzene diminish $\sim 1\%$ per degree increase in T_c near 300 K, as illustrated in Figure 2. Therefore, the most accurate determinations require the use of a cold-wall temperature that matches that used to measure the D_T values of the constituent homopolymers.

Theory

In ThFFF, the fundamental retention parameter λ is related to the temperature drop across the channel ΔT and the transport coefficients of the polymer-carrier liquid system by

$$\lambda = \frac{D}{D_T \Delta T} \tag{1}$$

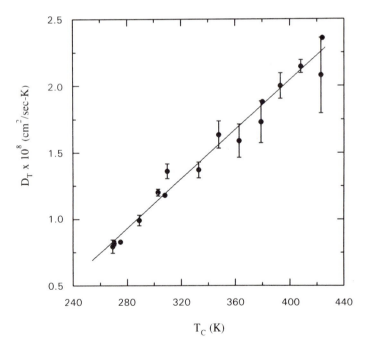

Figure 2. *Dependence of the thermal diffusion coefficient D_T on cold-wall temperature (T_c) for polystyrene in ethylbenzene. The data was taken from reference 12.*

where D is the ordinary diffusion coefficient. In most FFF subtechniques, λ is related to the volume of liquid required to elute a polymer component V_r by the following relation

$$R = V^o/V_r = 6\lambda[\coth{(\tfrac{1}{2}\lambda)} - 2\lambda] \tag{2}$$

Here R is termed the retention ratio and V^o is the void volume, that is, the volume required to elute a nonretained component. In ThFFF, corrections to equation 2 must be made to account for the viscosity gradient in the channel, which arises from the temperature gradient. The details of these corrections are lengthy and can be found in reference 13.

The diffusion coefficient can be related to the intrinsic viscosity $[\eta]$ and molecular weight of the polymer by (14)

$$D = \frac{kT}{6\pi\eta_o}\left(\frac{10\pi N_A}{3[\eta]M_V}\right)^{1/3} \tag{3}$$

where k is Boltzman's constant, T is temperature, N_A is Avogadro's number, η_o is the solvent viscosity, and M_V is the viscosity-average molecular weight. For a copolymer containing two components A and B, the thermal diffusion coefficient of the copolymer can be expressed as

$$D_T^{\text{copolymer}} = D_T^B + \frac{X_A}{100}(D_T^A - D_T^B) \tag{4}$$

Here, D_T^A and D_T^B are the thermal diffusion coefficients of homopolymers composed of components A and B, respectively, and X_A is the copolymer composition in mol% of component A. Substituting equations 3 and 4 into equation 1 and rearranging yields

$$\frac{1}{\lambda} = \frac{6\pi\eta_o\Delta T}{kT_{cg}}\left(\frac{3[\eta]M_V}{10\pi N_A}\right)^{1/3}\cdot(D_T^B + X_A\Delta D_T/100) \tag{5}$$

where ΔD_T has been substituted for $(D_T^A - D_T^B)$ and T_{cg} is the temperature in the channel at the center of gravity of the polymer distribution. T_{cg} is related to T_c by

$$T_{cg} = T_c + \lambda\Delta T \tag{6}$$

In characterizing a copolymer sample, we are specifically interested in two parameters contained in equation 5, namely M_V and X_A. The remaining parameters can either be found in handbooks or measured separately. Thus, parameter λ is calculated from V_r using the modified form of equation 2. D_T^A and D_T^B are known for several polymer-carrier liquid systems or they can be measured experimentally using ThFFF (9). Δ_T is set by the user and η_o (at T_{cg}) can be found in a handbook.

Finally, $[\eta]$ can be measured separately, leaving M_V and X_A as the two remaining unknown parameters. By obtaining viscosity and ThFFF retention data in two solvents, we establish two equations in the form of equation 5, which can be solved simultaneously to yield M_V and X_A

$$M_V = K\left[\frac{b_1 c_1/\lambda_2 - b_2 c_2/\lambda_1}{b_1 c_1 a_2 c_2 - b_2 c_2 a_1 c_1}\right]^3 \tag{7}$$

$$X_A = 100 \cdot \frac{a_2 c_2/\lambda_1 - a_1 c_1/\lambda_2}{b_1 c_1/\lambda_2 - b_2 c_2/\lambda_1} \tag{8}$$

where

$$a_i = D_T{}^B \qquad \text{(in solvent } i\text{)} \tag{9a}$$

$$b_i = \Delta D_T \qquad \text{(in solvent } i\text{)} \tag{9b}$$

$$c_i = \eta_o \Delta T [\eta]_i{}^{1/3}/T_{cg} \qquad \text{(in solvent } i\text{)} \tag{9c}$$

$$K = \left(\frac{6\pi}{k}\right)\left(\frac{3}{10\pi N_A}\right)^{1/3} \tag{9d}$$

Here the subscript i refers to the solvent, whereas the superscript (A or B) refers to the component homopolymer. For example, a_1 is the thermal diffusion coefficient of a homopolymer consisting of component B in solvent 1. Parameters $[\eta]_i$ and λ_i are the intrinsic viscosity and retention parameter measured on the copolymer in solvent i; T_{cg} in equation 9c is the temperature at the center of gravity of the retained polymer zone in solvent i, while η_o is the viscosity of solvent i at T_{cg}. Equations 7 through 9 are applicable to copolymers with only two components; similar equations could be derived for n-component copolymers, in which case M_V and X_A are determined from retention and viscosity data in n separate solvents.

The method outlined above is strictly applicable only to monodisperse copolymer samples, unless a viscometer is used for detection of the ThFFF elution profile. This arises because D in equation 3 is an average diffusion coefficient corresponding to a molecular weight M_V, which is not necessarily equal to the molecular weight at the ThFFF peak maximum. With a viscometer as the ThFFF detector, this limitation could be overcome by using the retention ratio corresponding to the center of gravity of the elution profile.

Experimental Details

The method was tested on a well-defined copolymer of styrene and isoprene, obtained from Exxon Chemical Company, Linden, NJ, and prepared by anionic polymerization to obtain a narrow molecular weight distribution.

The average molecular weight of the copolymer was determined by light scattering to be 38,000; the nominal composition is 51 mole percent styrene and 49 mole percent isoprene, determined by NMR.

Our ThFFF channel is similar to the model T100 Thermal Fractionator (FFFractionation, Inc., Salt Lake City, UT), with a channel thickness of 0.10 mm. When the carrier liquid was tetrahydrofuran (THF) or cyclo-hexane, a UV monitor set at 254 nm was used for sample detection; when toluene was the carrier liquid, a refractive index monitor was used instead. The temperature difference was 60.0 K and the cold wall temperature was 298.2 K. Intrinsic viscosities were measured with a CannonFenske ASTM-25 viscometer obtained from Fisher Scientific (Santa Clara, CA). Viscosities were measured in a thermostated temperature bath set at T_{cg}. All solvents were high-performance liquid chromatography grade.

Values of D_T^A and D_T^B were measured previously (5, 9), with relative standard errors that were typically 2%. Values of η_0 at T_{cg}, used in equation 9c, were obtained from reference 15.

Results and Discussion

Retention and viscosity of the copolymer were first measured in toluene and THF. A typical viscosity plot is illustrated in Figure 3. Here, the

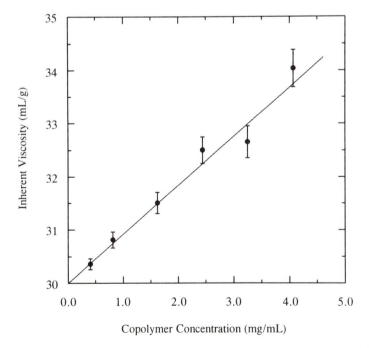

Figure 3.   Inherent viscosity versus concentration for test copolymer in THF. The intercept at zero concentration yields an intrinsic viscosity of 30.0 ± 0.3 mL/g.

inherent viscosity is plotted as a function of polymer concentration; the intercept is the intrinsic viscosity. In measuring ThFFF retention, a relatively high field strength (D_T 60 K) was required. Unfortunately, this resulted in a cold wall temperature (298.2 K) that was 4° higher than that used to obtain values of D_T^A and D_T^B, so we can expect a systematic error (evaluated below) in our results.

Two sets of experimental values are reported in Table I. The first set does not adjust the values of D_T^A and D_T^B to account for the higher cold wall temperature used in this work. In fact, such adjustments cannot be made with certainty because the effect of temperature on thermal diffusion in THF or toluene has not been studied. We can only make an approximate adjustment based on the temperature dependence of D_T observed for polystyrene in ethylbenzene. When this approximation is used, we get the second set of experimental results reported in Table I. Although the estimated composition differs from the nominal value by 3%, the discrepancy is within the uncertainty of the nominal value. However, the unadjusted molecular weight is biased by 16%. We report errors in composition in absolute terms, whereas errors in molecular weight are reported as relative errors.) We approximate the effect of a 4 K increase in cold wall temperature by increasing the values of D_T^A and D_T^B by 4%. This adjustment is based on the increase in D_T of 1% per degree for polystyrene in ethylbenzene. Although it does not affect the computed composition, the temperature adjustment moves the estimated molecular weight closer to the reported value, so that only a 3% discrepancy remains. Although a confidence level cannot be assigned

Table Ia.   Summary of Parameters in Toluene and THF

Parameters	Toluene	THF
D_T^A ($\times 10^8$ cm^2/s-K)	10.3	9.7
D_T^B ($\times 10^8$ cm^2/s-K)	6.9	5.4
λ	0.178	0.218
T_{cg} (K)	308.9	311.3
η_o (cP)	0.488	0.390
$[\eta]$ (mL/g)	17.5 ± 0.2	30.0 ± 0.5

Table Ib.   Summary of Results

Results	M_V (g/mol)	X_A (mol%)
Nominal	38,000	51
Experimental		
Unadjusted	44,000	54
D_T Adjusted to T_c	39,000	54

NOTES: A is polystyrene and B is polyisoprene.

to this revised estimate of molecular weight, we can be encouraged by the fact that the adjustment moves the estimated value in the right direction (i.e., the discrepancy in molecular weight becomes smaller rather than larger).

To further check the consistency of the method, we compare the copolymer's diffusion coefficient in THF, calculated from equation 3, with an independent measurement. Using our experimentally determined values of M_V and $[\eta]$, equation 3 predicts a D value of 9.85×10^{-7} cm^2/s; this is in good agreement with the value of 9.79×10^{-7} cm^2/s measured independently using NMR by workers at Exxon. (A similar comparison in toluene cannot be made because an independent value of D in toluene is not available.) A treatment of the propagation of errors is summarized in Table II. Here, uncertainties in M_V and X_A propagated by estimated uncertainties in the independent variables are listed. The largest uncertainty comes from the D_T values, but accurate retention data are also critical. For example, a 2% uncertainty in retention parameter λ translates to a 9–10% uncertainty in M_V and a 5–8% uncertainty in X_A.

Statistically, the numbers in Table II represent an upper limit to the uncertainties produced by random error because multiple sources of random error tend to cancel one another. Systematic errors, on the other hand, must be considered separately. In general, systematic errors are less critical to accuracy than random error due to the mathematical form of equations 7 and 8. Both equations are comprised of difference terms, where the two components of each difference term are composed of similar physical parameters.

Table II.   Summary of Uncertainties in Composition and Molecular Weight Propagated from the Independent Variables

Independent Variable	Error	Uncertainty in M_V (%)	Uncertainty in X_A (%)
ΔT	0.5 K	−1.5	0
T_C	0.2 K	0.8	0.1
$D_T^{A,1}$	3%	−16.3	9.8
$D_T^{A,2}$	2%	6.7	−5.5
$D_T^{B,1}$	2%	−5.9	3.4
$D_T^{B,2}$	2%	4.0	−3.3
All D_T values	4%	−13.2	0
λ_1	2%	−8.7	5.4
λ_2	2%	10.4	−8.1
$[\eta]_1$	2%	−5.4	3.1
$[\eta]_2$	2%	3.7	−3.0

NOTES: A is polystyrene, B is polyisoprene, 1 is toluene, and 2 is THF.

When a physical parameter in one component of a difference term is biased, the analogous parameter in the other component is similarly biased; as a result, the bias is partially canceled in the difference term. For example, when D_T values are biased by 4% (as expected from the mismatch in cold wall temperatures), the error in X_A is negligible, whereas the error in M_V is less than that caused by a similar error in $D_T^{A,1}$ alone (see Table II).

In this work, M_V was found to be more sensitive than X_A to systematic errors in D_T. This can be expected in other polymer-solvent systems as well. Thus, a bias in D_T values are manifested in the a_i terms of equations 7 and 8, yielding a relative error in the denominator of equation 7 that is of the same order of magnitude as the relative error induced in the numerator of equation 8. However, the cubic form of equation 7 results in more severe consequences for M_V compared with X_A.

A more detailed consideration of errors in retention and viscosity is illustrated with several plots in Figure 4. Each plot considers the error in one experimental measurement (viscosity or retention), and its effect on one of the two determinations (molecular weight or composition). The plots are three-dimensional because the consequence of errors in each measurement are examined in both solvents simultaneously. For example, consider the error in molecular weight caused by errors in the

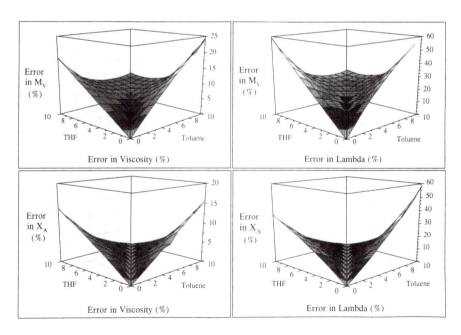

Figure 4.   Plots of error in M_V and X_A when retention and viscosity is measured in toluene and THF.

measurement of viscosity. Systematic errors are represented in the diagonal running from front to back, where errors are of similar magnitude in both solvents. Random errors are represented in the wings of the plot (i.e., off the diagonal running from front to back), where the error in one solvent dominates. It is clear from these plots that systematic errors are less critical to accuracy than random errors. In fact, a systematic error in either viscosity or retention has a negligible effect on the determination of composition.

The extreme sensitivity of the method to random errors in the retention parameter λ is apparent in Figure 4. Fortunately, we can obtain retention parameters with a precision of 1%. However, to avoid systematic errors, the corrected form of equation 2 (*see* reference 13) should be used to obtain values of λ from experimental measurements of R.

The mathematical form of equations 7 and 8 suggest that errors will diminish as differences in the magnitude of thermal diffusion between homopolymers and between solvents increase. Because thermal diffusion is comparable in toluene and THF but significantly different (weaker) in cyclohexane, we substituted cyclohexane for THF and reanalyzed the copolymer. The results are displayed in Table III. The estimated composition is 56 mole percent styrene in this solvent pair, which is still within the uncertainty of the nominal value. The estimated molecular weight remains 13% high. However, with a 4% adjustment in D_T values, the estimated molecular weight matches the nominal value.

A treatment of the propagation of errors in the toluene–cyclohexane system is summarized in Table IV and Figure 5. In this solvent pair, the

Table IIIa.   Summary of Parameters
in Toluene and Cyclohexane

Parameters	Toluene	Cyclohexane
D_T^A ($\times 10^8$ cm^2/s-K)	10.3	4.4
D_T^B ($\times 10^8$ cm^2/s-K)	6.9	0.8
λ	0.178	0.43
T_{cg} (K)	308.9	324.0
η_o (cP)	0.488	0.610
$[\eta]$ (mL/g)	17.5	22.0

Table IIIb.   Summary of Results

Results	M_V (g/mol)	X_A (mol%)
Nominal	38,000	51
Experimental		
Unadjusted	43,000	56
D_T adjusted to T_c	38,000	56

NOTES: A is polystyrene and B is polyisoprene.

Table IV.   Summary of Uncertainties in Composition and Molecular Weight
Propagated from the Independent Variables

Independent Variable	Error	Uncertainty in M_V (%)	Uncertainty in Composition (%)
ΔT	0.5 K	−0.7	0.1
T_C	0.2 K	0.8	0.1
$D_T^{A,1}$	3%	−7.6	1.9
$D_T^{A,3}$	2%	3.1	−2.5
$D_T^{B,1}$	2%	−4.6	1.2
$D_T^{B,3}$	2%	0.6	−0.4
All D_T values	4%	−5.8	0
λ_1	2%	−4.3	0.9
λ_3	2%	1.6	−0.8
$[\eta]_1$	2%	−3.1	0.8
$[\eta]_3$	2%	0.7	−0.6

NOTES: A is polystyrene, B is polyisoprene, 1 is toluene, and 3 is cyclohexane.

effect of random error on calculations of both molecular weight and composition is less dramatic, although errors in D_T^A and D_T^B remain critical, particularly for polystyrene in toluene. However, the sensitivity of molecular weight to errors in the measurement of viscosity in toluene is cut nearly in half, and in cyclohexane it is reduced by a factor of five compared with THF; the effect on composition is reduced by a factor of four in both solvents. An even more dramatic reduction occurs in the sensitivity of the results to errors in retention. In toluene, the effect on molecular weight is reduced by a factor of two, whereas the effect on composition is reduced sixfold. In cyclohexane, the sensitivity of molecular weight to retention error is reduced by a factor of seven compared with THF, whereas a 10-fold reduction is seen in the effect on composition. In general, measurement errors are much less costly when toluene and cyclohexane are used as compared with toluene and THF.

Although it is clear that random errors are less problematic when cyclohexane is substituted for THF, the effect of systematic error remains the same. Because we found little change in agreement between calculated and nominal values when we switched solvent pairs, it is likely that systematic error is responsible for the discrepancies. The most probable source of discrepancy is systematic errors in D_T^A and D_T^B due to the higher cold wall temperature used in this work.

Conclusions

The molecular weight and chemical composition of copolymers can be determined by ThFFF and viscometry when either the primary or sec-

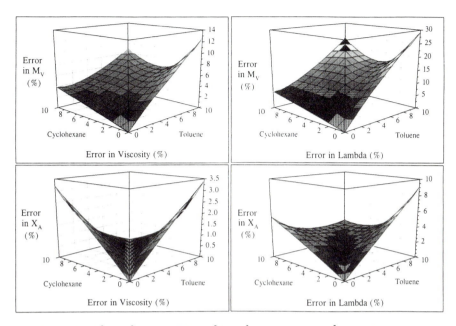

Figure 5.   *Plots of error in* M_V *and* X_A *when retention and viscosity is measured in toluene and cyclohexane.*

ondary structure of the copolymer is randomized. Although the primary structure is not random in block copolymers, previous studies indicate that the secondary structure of linear-block copolymers is random in nonselective solvents, that is, solvents that are equally good for all components. However, the method requires one solvent for each component, and it may be difficult to find multiple nonselective solvents, even for copolymers with only two components. In this case, it may be possible to use the temperature dependence of viscosity and ThFFF retention by using a single (nonselective) solvent at two different temperatures, because thermal diffusion and viscosity are both functions of temperature. Of course, such a method will only be reliable if the temperature dependence of viscosity and thermal diffusion is different for each component. For example, D_T^A and D_T^B cannot both increase by the same factor when the temperature is increased a given amount, or the orthogonality of the data is lost. Thus, more work is required to determine the general feasibility of the method to block copolymers, although in principle the applicability is there.

The accuracy of the method is quite sensitive to errors in the D_T values of the homopolymers. Fortunately, D_T values are independent of molecular weight and can be determined with precision using ThFFF. However, the relatively strong temperature dependence of D_T requires the use of a consistent cold wall temperature. Accurate retention values are also critical. Therefore, the extra-column volume should be accounted for in calculating the retention ratio R. Furthermore, the nonisoviscous corrections to retention theory should be used in obtaining λ from R when maximum accuracy is desired.

Acknowledgments

This work was funded by Idaho EPScOR and National Science Foundation grant OSR–9350539. Thanks go to Louise Wheeler and Exxon Chemical Company for the copolymer sample.

References

1. Myers, M. N.; Chen, P.; Giddings, J. C. In *Chromatography of Polymers: Characterization by SEC and FFF*; Provder, T., Ed.; ACS Symposium Series 521; American Chemical Society: Washington, DC, 1993; Chapter 4, pp 47–62.
2. Schimpf, M. E. *J. Chromatogr.* 1990, *517*, 405.
3. Schimpf, M. E.; Williams, P. S.; Giddings, J. C. *J. Appl. Polym. Sci.* 1989, 37, 2059.
4. Giddings, J. C.; Kumar, V.; Williams, P. S.; Myers, M. N. In *Polymer Characterization: Physical Property, Spectroscopic, and Chromatographic Methods*; Craver, C. D.; Provder, T., Eds.; Advances in Chemistry 227; American Chemical Society: Washington, DC, 1990; Chapter 1, pp 3–21.
5. Schimpf, M. E.; Rue, C. A.; Mercer, G. D.; Wheeler, L. M.; Romeo, P. F. *J. Coatings Technol.* 1993, *65*, 51.
6. Giddings, J. C.; Caldwell, K. D.; Myers, M. N. *Macromolecules* 1976, *9*, 106.
7. Gunderson, J. J.; Giddings, J. C. *Macromolecules* 1986, *19*, 2618.
8. Schimpf, M. E.; Giddings, J. C. *Macromolecules* 1987, *20*, 1561.
9. Schimpf, M. E.; Giddings, J. C. *J. Polym. Sci., Polym. Phys. Ed.* 1989, *20*, 1317.
10. Kirkland, J. J.; Rementer, W. *Anal. Chem.* 1992, *64*, 904.
11. Schimpf, M. E.; Giddings, J. C. *J. Polym. Sci., Polym. Phys. Ed.* 1990, *28*, 2673.
12. Brimhall, S. L.; Myers, M. N.; Caldwell, K. D.; Giddings, J. C. *J. Polym. Sci., Polym. Phys. Ed.* 1985, *23*, 2443.
13. Gunderson, J. J.; Giddings, J. C. *Sep. Sci. Technol.* 1984, *19*, 667.
14. Rudin, A.; Johnston, H. K. *J. Polymer Sci., Part B*, 1971, *9*, 55.
15. Landolt, H. H. In *Zahlen Werte und Functionen aus Physik, Chemie, Astronomie, Geophysik und Technik*; Springer: Berlin, Germany, 1969; Vol. 2, Part 5, Sec. a.

RECEIVED for review January 6, 1994. ACCEPTED revised manuscript July 19, 1994.

Compositional Heterogeneity of Copolymers by Coupled Techniques with Chromatographic Columns and Multidetectors

John V. Dawkins

Department of Chemistry, Loughborough University of Technology, Loughborough, Leicestershire LE11 3TU, England

For copolymers having composition and molar mass distributions, it is shown that characterization with one or more concentration detectors on-line to a chromatographic system based on size-exclusion chromatography (SEC) produces average composition data. For more detailed information on composition heterogeneity, two approaches are reviewed. An SEC method involving on-line concentration detection together with on-line low-angle laser light scattering is described to demonstrate how heterogeneity parameters permit a distinction between block copolymers and polymer blends. Coupled column chromatography with two chromatographic systems in which fractions from an SEC column are injected into a second column containing a polymer-based packing where retention is determined by nonexclusion mechanisms is described.

SIZE-EXCLUSION CHROMATOGRAPHY (SEC) is well established as a technique for determining the molar mass distribution (MMD) of homopolymers. Heterogeneous copolymers contain distributions in both molar mass (M) and copolymer composition. Copolymer characterization based on SEC is often performed with on-line selective concentration detectors (*1, 2*). For heterogeneous copolymers this SEC-based method is only capable of producing average composition data across a chromatogram, because copolymer chains having the same molecular size in solution will have variations in molar mass and composition (*3*).

0065–2393/95/0247–0197$12.00/0

Off-line light scattering has been developed to determine data for
M and compositional heterogeneity for copolymers (4-7). The parameters used to quantify the compositional heterogeneity of a copolymer
sample, as determined from light-scattering data, are P, representing
the effect of M on compositional heterogeneity, and Q, which represents
the overall compositional drift. Therefore, addition of on-line low-angle
laser light-scattering (LALLS) detection to an SEC system with dual
concentration detection can permit for some types of copolymers the
calculation of compositional heterogeneity at each elution volume together with overall heterogeneity parameters (8, 9). When copolymers
contain composition heterogeneities, some types of cross-fractionation
procedure, involving separating by composition fractions previously
separated by size, can be attempted, but the experimental work involving
transfers between techniques is quite time-consuming. In coupled column chromatography (CCC), on-line transfer can be automated, and
Balke and Patel (10-13) demonstrated copolymer separations with two
chromatographic systems in which copolymer is separated first by SEC
and second by nonexclusion mechanisms.

Here, some aspects of copolymer characterisation by SEC with coupling are reviewed considering, first, concentration detectors with
LALLS detection and, second, concentration detection with on-line
transfer to an interactive column system in CCC. Investigations of CCC
indicate how nonexclusion separations dependent on copolymer composition in the second column can be influenced by choice of stationary
and mobile phases (14, 15). The examples of statistical and block copolymers are selected to illustrate not only heterogeneity within copolymer chains but also homopolymer contamination within copolymer
samples. The presence of residual homopolymer is important to the
production of comb graft copolymers by grafting-on and grafting-through
processes (16).

Experimental

Chromatographic data for copolymers obtained by SEC with on-line dual
concentration detectors was gathered with a gel permeation chromatograph
(model 301, Waters Associates, Milford, MA) with refractive index (RI)
(thermostatted at 298 K) and UV (254 nm) detection (17). Elutions were
performed with tetrahydrofuran (THF) (distilled before use) at a flow rate
of 1 cm³ min⁻¹ at room temperature. A series arrangement of four SEC
columns (Styragel, Waters Associates) was used. Solution concentrations
were in the range of 0.1-0.3% (wt/vol). Calibrations of detector proportionality constants were established according to methodology described
previously (17).

Details of the chromatographic system with on-line LALLS were described previously (8, 9). After a series arrangement of four SEC columns
(300 × 7 mm PLgeL, 10 μm, Polymer Laboratories, Church Stretton,

England), the on-line detectors were in sequence light scattering (model KMX-6, Chromatix Thermo Separation Products, Riviera Beach, FL), infrared (model 1A, Wilks-Miran), and RI (model R401, Waters Associates). Separations were performed with tetrachloroethylene as eluent at 353 K. Solution concentrations were 5 mg/cm^3 and toluene (0.1%) was added as internal marker.

The CCC instrumentation consisted of two independent chromatographic systems joined together via a switching valve (*14, 15*). System one contained in sequence an SEC column (300 × 7 mm, mixed PLgeL, 10 μm), six port-switching valve (model 7010, Rheodyne, Cotati, CA), and RI detector (type 98.00, Knauer, supplied by Polymer Laboratories, Church Stretton, England). System two had pump and column linked through the same switching valve and contained in sequence a single column of either mixed PLgeL, 10 μm, or PL Aquagel P3 type, 10 μm (both 300 × 7 mm), UV detector (Pye Unicam), and RI detector (type 98.00, Knauer, supplied by Polymer Laboratories, Church Stretton, England). All separations were performed at ambient temperature. THF was always used as eluent in system one. Isocratic elutions were performed with the second system with mixtures of either THF–heptane (HEP) or THF–isopropanol (IP) as mobile phase. Solution concentrations in THF injected into the first column system were 0.4% (wt/vol).

Polymers and copolymers were laboratory-prepared samples. Samples W4 and W7 of the diblock copolymer AB poly(styrene-*b*-tetramethylene oxide) (PS–PT) were synthesized by producing a polystyrene prepolymer whose terminal group was transformed to a macroinitiator for the polymerization of THF. Samples B13 and B16 of the diblock copolymer AB poly[styrene-*b*-(dimethyl siloxane)] (PS–PDMS) were prepared by sequential anionic polymerization. Samples of statistical copolymers of styrene and *n*-butyl methacrylate (PSBMA) were produced by radical copolymerization. Details of synthetic and characterization methods have been reported elsewhere (*15, 17–19*).

Results and Discussion

Composition Drift.   Determinations of copolymer composition distribution by SEC with dual UV and RI detectors were developed by several researchers (*1, 20–22*). To exemplify that this methodology may only be capable of producing average composition data across a chromatogram as a function of retention volume V, results for samples of the diblock copolymer PS–PT are presented. The block lengths in samples of PS–PT were chosen such that the SEC peak for copolymer was well resolved from the PS prepolymer peak.

The response $h_{UV}(V)$ of the UV detector as a function of V depends only on the weight w_S of styrene units in PS–PT, whereas the response $h_{RI}(V)$ of the RI detector depends on both w_S and weight w_T of tetramethylene oxide units in the copolymer. The detector responses are given by

$$h_{UV}(V) = K_S w_S \tag{1}$$

$$h_{RI}(V) = K_C(w_S + w_T) \tag{2}$$

where K_S is a proportionality constant dependent on the UV extinction coefficient for styrene units in the copolymer, and K_C is a proportionality constant related to the RI increment of the PS–PT diblock copolymer in the SEC eluent. This increment is usually assumed to be represented in terms of the values for the corresponding homopolymers by a linear equation, so that K_C is given in terms of weight fractions by

$$K_C = W_S K_A + (1 - W_S) K_B \qquad (3)$$

where K_A and K_B are the refractometer proportionality constants for PS and PT, respectively. The weight fraction of styrene units in the copolymer is

$$W_S = w_S/(w_S + w_T) \qquad (4)$$

It follows that substitution of equations 1 and 2 into equation 4, and elimination of K_C with equation 3, gives after rearrangement an expression for W_S as a function of V

$$W_S(V) = \frac{K_B h_{uv}(V)/h_{RI}(V)}{K_S - (K_A - K_B) h_{uv}(V)/h_{RI}(V)} \qquad (5)$$

Because of the results to be presented to illustrate composition drift, we prefer to define the average weight fraction of styrene $\overline{W}_S(V)$ in equation 5. Determinations of K_S, K_A, and K_B were reported previously (17). It was demonstrated that for the range of molar masses studied there was no dependence of refractometer proportionality constants on chain length or end group structure.

Copolymer composition data for SEC peaks corresponding to two samples of PS–PT diblock copolymer are displayed in Figure 1. Sample W4 appears to be close to monodisperse both in terms of MMD, polydispersity computed to be 1.04 for chains eluting over the range 21.0–26.5 counts, and in terms of composition distribution, because the increase in $\overline{W}_S(V)$ above 0.12 at the peak of the chromatogram corresponds to the low molar mass tail of the SEC chromatogram where the accuracy of the dual detector method will decrease. On the other hand, sample W7 is much more polydisperse both in terms of MMD, polydispersity computed to be 1.65 for chains eluting over the range 19.5–27.5 counts, and in terms of composition distribution, because there is considerable increase in $\overline{W}_S(V)$ across the chromatogram from the value of 0.1 at the peak. These very different composition distributions for samples W4 and W7 may be explained by the type of chemistry used in the transformation reaction to produce a macroinitiator for the polymerization of THF in formation of PS–PT diblock copolymer (17).

At the ith elution volume interval in the elution of copolymer by SEC, a detector having a cell volume ΔV will provide a response cor-

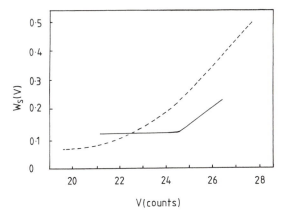

Figure 1. Dependence of composition of styrene units in PS–PT diblock copolymers on retention volume. ——, sample W4; – – –, sample W7.

responding to chains with almost identical sizes in solution in the cell, as judged by hydrodynamic volume as universal calibration parameter. If there is a drift of composition across a chromatogram, as exemplified by sample W7 in Figure 1, then it can be expected that there has to be a compositional heterogeneity for chains of almost identical sizes in the detector cell volume ΔV at a particular elution time. From universal calibration considerations (17), the experimental SEC calibrations for PS and PT homopolymers are related at a given elution volume by

$$\log M_{PT} - \log M_{PS} = \log 0.55 \tag{6}$$

Therefore, heterogeneous copolymer chains in a detector cell volume ΔV will have different molar masses. It follows that for a definite hydrodynamic volume of chains in solution these chains may be constituted by a range of structures with variations in block lengths and composition. Consequently, for heterogeneous copolymers, SEC with concentration detectors is only capable of producing average composition data, and more detailed studies of compositional heterogeneity require additional characterization methodology, that is, by LALLS for some polymer types or by cross-fractionation.

LALLS Detection. Consideration of the treatment of light scattering for heterogeneous copolymers (6) permits the dependence of the apparent molar mass M_i^* at the ith elution volume interval in a SEC–on-line LALLS experiment to be represented in terms of P_i and Q_i by

$$M_i^* = M_{wi} + 2P_i\left(\frac{\nu_A - \nu_B}{\nu_i}\right) + Q_i\left(\frac{\nu_A - \nu_B}{\nu_i}\right)^2 \tag{7}$$

where ν_i is the RI increment for all the components (having weight average molar mass M_W) at the ith elution volume interval. If one assumes a linear relation between RI increment and copolymer composition W_i (determined for chains at the ith elution volume interval from peak responses from on-line concentration detectors), it is easy to calculate ν_i from the homopolymer RI increments ν_A and ν_B and the measured W_i value by analogy with the method defined by equation 3. The apparent molar mass M_i^* is determined with equation 8,

$$K^*\nu_i^2 c_i / \overline{R_{\theta i}} = (1/M_i^*)\text{ERROR}^*) + 2A_{2i}c_i \qquad (8)$$

which defines the excess Rayleigh factor $\overline{R_{\theta i}}$ due to scattering from solute alone (concentration c_i) at the ith elution volume interval. In this equation the term containing the second virial coefficient A_{2i} can be neglected for SEC experiments at low values of c_i, and K^* contains the usual constants in light scattering.

Diblock copolymers of PS–PDMS were chosen for study because PS and PDMS homopolymers in good solvents have the same molar mass calibrations in SEC (23). For PS and PDMS homopolymers in tetrachloroethylene, it can be shown from data for intrinsic viscosity that the Mark–Houwink exponent for both of these polymers is near 0.8 (9). Equations for universal calibration (24) indicate that an M(PS–PDMS) diblock copolymer calibration should therefore follow that for the corresponding homopolymers. Consequently, there should be a narrow range of masses at each elution volume, so that the term containing P_i in equation 7 can be ignored and M_{wi} can be replaced by M_i, giving

$$M_i^* = M_i + Q_i\left(\frac{\nu_A - \nu_B}{\nu_i}\right)^2 \qquad (9)$$

Because M_i is known from SEC calibrations with PS and PDMS homopolymer standards and because M_i^* can be determined from on-line LALLS and concentration detectors with the conventional light-scattering equation containing the excess Rayleigh factor (equation 8), sufficient information is available to compute Q_i across a chromatogram. These values can then be averaged to obtain the heterogeneity parameter Q for the overall sample. To facilitate comparisons among samples, it is convenient to use another heterogeneity parameter, H, defined as

$$H = Q/Q_{\text{max}} \qquad (10)$$

where Q_{max} is the value obtained for a blend of two homopolymers. The range of H values is from zero (homogeneous sample) to unity (maximum heterogeneity, i.e., a blend).

Table I. Molar Mass and Heterogeneity Data
for a Blend and a Diblock Copolymer

Characterization Parameter	Blend of PS + PDMS Homopolymers (Sample 4)		Diblock Copolymer PS–PDMS (Sample B16)	
	Expected Value	Value SEC/LALLS	Expected Value	Value SEC/LALLS
PS (or styrene) content (%)	75.9[a]	75.5	76.1[b]	74.8
M(PS)	43,600[c]	44,800	43,600[c]	39,700
M(PDMS)	35,800[c]	43,500	13,400[d]	13,400
H	1.00	1.06	0	0.25

[a] Blend composition weighed out.
[b] Calculated from Si analysis.
[c] Independent SEC characterization of homopolymers.
[d] Computed from b and c.

Representative results for a blend and a diblock copolymer are shown in Table I. It is evident for both samples that the on-line infrared and RI concentration detectors provide excellent estimates of overall composition, component molar masses, and the heterogeneity parameter H. Data for blends in Table I were obtained to assess the proposed methods. Further evidence for the capability of the SEC–LALLS procedure can be seen in Figure 2, which shows plots of data for W_S and H computed across the MMD for blend sample 4. In this figure the expected results

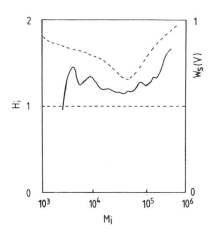

Figure 2. Variation of composition of PS and compositional heterogeneity across MMD for PS–PDMS blend sample 4. - - -, composition $W_S(V)$; ——, heterogeneity parameter H_i.

for W_S and H from Table I are obtained for the M_i range from 10^4 to 10^5 g mol^{-1}. This indicates that possible errors due to volume offset among the three detectors have been minimized by analysis of marker peaks. The greatest errors in the plots in Figure 2 are at the tails of the distribution, which are partly due to the PS-rich blend, the greater polydispersity of the PS component than PDMS (4 and 2, respectively) and possible distortion of tails by band-broadening.

The results for copolymer sample B16 in Table I suggest that it is a good homogeneous diblock copolymer with minimal levels of contaminating homopolymers. This is confirmed by plots of data for W_S and H computed from outputs from the three on-line detectors across the MMD and displayed in Figure 3. The weight fraction of styrene is close to 0.75 over about a decade of M, and it is only at the low molar mass tail of the distribution, where the multidetector approach will have lowest accuracy, that W_S decreases significantly below the mean value quoted in Table I. Characterization results for a much more polydisperse sample B13, in terms of range of molar masses, are also shown in Figure 3. This sample, although synthesized by a sequential monomer addition process to produce a diblock copolymer, exhibits a fluctuating trend in terms of

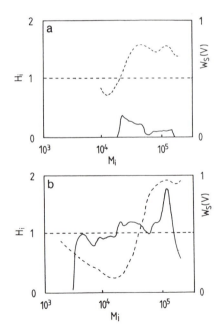

Figure 3. *Variation of composition of styrene units and compositional heterogeneity across MMD for PS–PDMS copolymer samples; (a) sample B16 and (b) sample B13. – – –, composition $W_S(V)$; ——, heterogeneity parameter H_i.*

W_S, indicating fractions having styrene-rich and styrene-deficient components. It is difficult to rationalize these composition data with the method of block copolymer synthesis, and it has to be considered that sample B13 contains substantial contamination by one or both homopolymers. The interpretation that this sample is largely a polydisperse blend of polymers rather than based on a copolymer is supported by the plot of H that lies in the range 0.80–1.06 across the peak of the MMD in Figure 3. This deduction would not have been possible by examination of average composition data alone without the application of light scattering to determine heterogeneity parameters. The overall information obtained from the three detectors enables molar masses of components to be determined, and for the two samples in Table I the good agreement between expected and SEC–LALLS results indicates that the methods proposed are reasonably accurate.

CCC.   A schematic diagram of the CCC system is shown in Figure 4 [this technique also has been named orthogonal chromatography (*10–13*)]. Cross-fractionation on the polymer solution injected into the SEC column was performed on the solution passing through the switching valve in the time interval 830–850 s. This fraction was separated by isocratic elutions with column two.

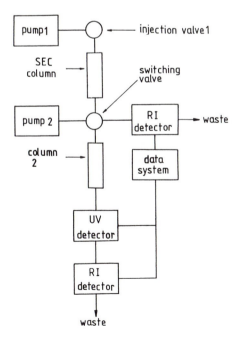

Figure 4.   Diagram of apparatus for CCC.

THF/ HEP   V(s)   800   1000

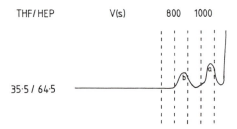

35·5 / 64·5

Figure 5.   Chromatogram obtained with column system 2 containing PLgeL PS packing for a mixture of homopolymers of PBMA (peak b) and PS (peak a) with isocratic elution of THF–HEP (35.5:64.5) as mobile phase.

Cross-fractionation in CCC requires the establishment of separation conditions for homopolymers for the second column system. A minimum requirement is to produce nonoverlapping chromatograms by identifying mobile-phase compositions for resolution of PS and PBMA homopolymer peaks. This separation is obtained by introducing HEP or IP, as nonsolvent component for PS, and resolution of homopolymer peaks for the PLgeL PS packing was obtained with mobile-phase compositions of 35.5: 64.5 (THF–HEP), as shown in Figure 5, and 55:45 (THF–IP), as shown in Figure 6. An increase in nonsolvent concentration in the mobile phase markedly shifts PS elution to longer retention times, whereas PBMA exhibits little or no change in elution volume. The behavior of PS is consistent with separations of PS in poor and theta solvents with cross-linked PS gels (25), that is, nonexclusion interaction mechanisms are similar for poor solvents that are both more polar (IP) and less polar

V(s)   800   1000

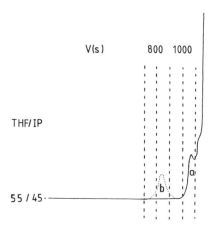

THF/ IP

55 / 45·

Figure 6.   Chromatograms obtained with column system 2 containing PLgeL PS packing for homopolymers of PBMA (peak b) and PS (peak a) with isocratic elution of THF–IP (55:45) as mobile phase.

(HEP) than PS. It is proposed that size exclusion continues to dominate separations of PBMA for the mobile phases given previously. For PL Aquagel P3 packing in the second column, resolution of homopolymer peaks was obtained with mobile-phase compositions of 55:45 (THF–HEP) and 30:70 (THF–IP). However, these results are not as easy to interpret as the observations for PLgeL because both PS and PBMA homopolymer peaks appear to be influenced by adsorption on PL Aquagel P3 as the HEP concentration is increased and because there is a low difference in peak retention volumes between these peaks for elutions with THF–IP, requiring a very high fraction of IP to achieve peak resolution.

Consequently, separation of PSBMA copolymers according to styrene composition ought to be possible in the second column system containing a PLgeL packing. Separations of three different PSBMA copolymers in mixtures with PBMA are shown in Figure 7. Chromatogram a shows the PBMA peak eluting first and the peak due to copolymer PSBMA8/2 merging with the solvent peak at 1200 s. As the styrene composition decreases, the copolymer is less retained exhibiting decreased V_R as shown by chromatograms b and c. Two peaks are observed in each case, due to PBMA with V_R near 900 s and copolymer eluting later. Therefore, this CCC method has potential not only for separating copolymers on the basis of composition but also for isolating residual homopolymers from copolymers. The latter problem is of relevance to the production of comb graft copolymers by grafting-on and grafting-through processes (Slark, A. T.; Azam, M.; Branch, M. G.; Dawkins, J. V., Loughborough University of Technology, Loughborough, United Kingdom, unpublished results.)

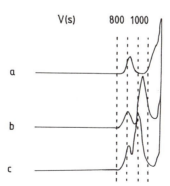

Figure 7.   Chromatograms obtained with column system 2 containing PLgeL PS packing with THF–HEP (composition 30:70) as mobile phase. (a) PBMA homopolymer and PSBMA copolymer (65 mol/% styrene); (b) PBMA homopolymer and PSBMA copolymer (51 mol/% styrene); and (c) PBMA homopolymer and PSBMA copolymer (36 mol/% styrene).

Conclusions

The results demonstrate that coupled chromatographic techniques with multiple detectors permit the determination of average composition data, heterogeneity parameters, and separations of homopolymers and copolymers. The methodology reviewed here enables a distinction to be made between copolymers and polymer blends.

Acknowledgments

I thank T. Dumelow, S. R. Holding, A. M. C. Montenegro, and G. Taylor for helpful discussions. Support for part of this work was kindly provided by the Polymer Supply and Characterization Centre at Rubber and Plastics Research Association and by the Science and Engineering Research Council.

References

1. Runyon, J. R.; Barnes, D. E.; Rudd, J. F.; Tung, L. H. *J. Appl. Polym. Sci.* **1969**, *13*, 2359.
2. Quivoron, C. In *Steric Exclusion Liquid Chromatography of Polymers*; Janca, J., Ed.; Marcel Dekker: New York, 1984; pp 213–280.
3. Hamielec, A. *Pure. Appl. Chem.* **1982**, *54*, 293.
4. Bushuk, W.; Bemoit, H. *Can. J. Chem.* **1956**, *36*, 1616.
5. Krause, S. *J. Phys. Chem.* **1961**, *65*, 1618.
6. Leng, M.; Benoit, H. *J. Polym. Sci.* **1962**, *57*, 263.
7. Benoit, H.; Froelich, D. In *Light Scattering from Polymer Solutions*; Huglin, M. B., Ed.; Academic: London, 1972; p 467.
8. Dumelow, T. Ph.D. Thesis, Loughborough University of Technology, Loughborough, United Kingdom, 1984.
9. Dumelow, T.; Holding, S. R.; Maisey, L. J.; Dawkins, J. V. *Polymer* **1986**, *27*, 1170.
10. Balke, S. T.; Patel, R. D. In *Size Exclusion Chromatography (GPC)*; Provder, T., Ed.; ACS Symposium Series 138; American Chemical Society: Washington, DC, 1980; pp 149–152.
11. Balke, S. T.; Patel, R. D. *J. Polym. Sci. B* **1980**, *18*, 453.
12. Balke, S. T.; Patel, R. D. In *Polymer Characterization—Spectroscopic, Chromatographic and Physical Instrumental Methods*; Craven, C., Ed.; ACS Advances in Chemistry 203; American Chemical Society: Washington, DC, 1983; pp 281–310.
13. Balke, S. T. In *Detection and Data Analysis in Size Exclusion Chromatography*; Provder, T., Ed.; ACS Symposium Series 352; American Chemical Society: Washington, DC, 1987; pp 59–77.
14. Montenegro, A. M. C. Ph.D. Thesis, Loughborough University of Technology, Loughborough, United Kingdom, 1986.
15. Dawkins, J. V.; Montenegro, A. M. C. *Br. Polym. J.* **1989**, *21*, 1.
16. Rempp, P.; Lutz, P. J. In *Comprehensive Polymer Science*; Allen, G.; Bevington, J. C., Eds.; Pergamon: Oxford, England, 1989.
17. Burgess, F. J.; Cunliffe, A. V.; Dawkins, J. V.; Richards, D. H. *Polymer* **1977**, *18*, 733.

18. Taylor, G. Ph.D. Thesis, Loughborough University of Technology, Lough-
    borough, United Kingdom, 1977.
19. Dawkins, J. V.; Taylor, G. *Polymer* **1979**, *20*, 559.
20. Adams, H. E. *Sep. Sci.* **1971**, *6*, 259.
21. Grubisic-Gallot, Z.; Picot, M.; Gramain, P. R.; Benoit, H. *J. Appl. Polym.
    Sci.* **1972**, *6*, 2931.
22. Harmon, D. J.; Folt, V. L. *Rubber Chem. Technol.* **1973**, *46*, 448.
23. Dawkins, J. V. *J. Macromol. Sci.-Phys.* B **1968**, *2*, 623.
24. Dawkins, J. V.; Guest, M. J.; Jeffs, G. M. F. *J. Liq. Chromatogr.* **1984**, *7*,
    1739.
25. Dawkins, J. V.; Hemming, M. *Makromol. Chem.* **1975**, *176*, 1795.

RECEIVED for review January 6, 1994. ACCEPTED revised manuscript July 21,
1994.

Size-Exclusion Chromatography and Nonexclusion Liquid Chromatography for Characterization of Styrene Copolymers

Sadao Mori

Department of Industrial Chemistry, Faculty of Engineering, Mie University, Tsu, Mie 514, Japan

A size-exclusion chromatography (SEC) dual detector system cannot give an accurate chemical composition distribution (CCD) for copolymers; therefore, nonexclusion liquid chromatography (NELC) is required. SEC–NELC can give both a molecular-weight distribution and a CCD for copolymers accurately and precisely. Several NELC techniques that separate copolymers according to composition are reviewed: liquid adsorption chromatography (LAC), high-performance precipitation liquid chromatography, normal- and reversed-phase chromatography, orthogonal chromatography, and LAC at the critical point. LAC with silica gel–chloroform (or 1,2-dichloroethane) + ethanol for the separation of styrene copolymers is explained in detail.

M$_{OST}$ SYNTHETIC POLYMERS (homopolymers) have a molecular-weight distribution (MWD) that can be determined by size exclusion chromatography (SEC). Although molecular-weight averages and the MWD of copolymers also can be determined by SEC, accurate information on these values is not always easy to obtain by SEC alone. SEC separates molecules by their hydrodynamic sizes in solution, not by their molecular weights. Copolymers have both a MWD and a chemical composition distribution (CCD); a fraction eluted at the same retention volume in SEC is a mixture of molecules having different compositions (and different molecular weights) but the same hydrodynamic size.

0065–2393/95/0247–0211$12.00/0

An SEC dual detector system is a well-known technique to determine the CCD of copolymers. As a combination of detectors, UV and refractive index detectors are commonly used. However, because SEC separates copolymers according to molecular sizes as described previously, the elution order of the copolymers is not proportional to composition difference, and components eluted in the same retention volume might have different compositions of the same size. Therefore, only the average composition at each retention volume can be detected by SEC with two detectors (1). If the CCD calculated from a dual detector system shows the increase, the decrease, or the fluctuation with retention volume, then the copolymer has chemical heterogeneity. Even if the CCD shows the sample copolymer to be homogeneous throughout the whole chromatogram, it is almost impossible to conclude that the copolymer is homogeneous.

Techniques such as SEC–LC (liquid chromatography other than the size exclusion separation mode) are required to characterize copolymers in accurate detail. Several techniques for nonexclusion liquid chromatography (NELC) to separate copolymers according to composition have been developed and reported within the past several years. These techniques can give the information on chemical heterogeneity of copolymers; thus, SEC–NELC is required to determine both distributions.

Techniques for NECL

NECL was first used to separate copolymers according to composition by Teramachi et al. in 1979 (2). Silica gel was the stationary phase, and styrene–methyl acrylate copolymers were separated by a gradient elution method with a combination of carbon tetrachloride–methyl acetate as the mobile phase. Styrene–methyl methacrylate (MMA) copolymers were separated by a dichloroethane–tetrahydrofuran (THF) gradient elution method (3). High-performance precipitation liquid chromatography (HPPLC) was developed by Glöckner et al. (4) and was applied to the separation of styrene–acrylonitrile copolymers. A mixture of THF and n-hexane was used as the mobile phase and gradient elution was performed in order of increasing THF content. Orthogonal chromatography was developed by Balke and Patel (5) and applied to the separation of styrene-n-butyl methacrylate copolymers. Polystyrene gel was used as the stationary phase and THF and a mixture of THF and n-heptane were the first and the second mobile phases, respectively.

These NELC techniques can be classified into five types. Each type requires gradient elution with two or more solvents. The first type is liquid adsorption chromatography (LAC) (Table I). In LAC, initial and final mobile phases should be good solvents for the sample copolymers. Silica gel is used as the stationary phase in most cases. The sample co-

Table I. Examples for LAC

Mobile Phase	Stationary Phase	Copolymer	Ref.
CCl$_4$/AcOMe	Silica gel	P(S–MA)	2
DCE/THF	Silica gel	P(S–MMA), P(S–EMA)	3
Toluene/MEK	Silica gel	Polyacrylates, polymethacrylates (homo- and copolymers)	6
CHCl$_3$/CHCl$_3$ + EtOH	Silica gel	P(S–methacrylates) P(S–acrylates)	7, 8 8
DCE/DCE + EtOH	Silica gel	P(S–methacrylates) P(S–acrylates) Polymethacrylates, P(EMA–PBMA)	9 10

NOTE: S, polystyrene; MA, poly(methyl acrylate); EMA, poly(ethyl methacrylate); BMA, poly(butyl methacrylate).

polymers injected into a column adsorb on the silica gel surface at the initial mobile phase. For example, the initial mobile phase in reference 3 was 3% THF in 1,2-dichloroethane (DCE) and in reference 6 was 2% methyl ethyl ketone (MEK) in toluene. All the sample copolymers prefer to adsorb on the surface of the stationary phase rather than elute from the column. Gradient elution increases the content of a displacer, which acts to decrease the adsorption power of the stationary phase (e.g., ethanol in references 7–10) or to increase the solubility of the copolymers (e.g., final mobile phase in reference 3 was 15% THF in DCE and in reference 6, 100% MEK). In this sense, both mobile and stationary phases play important roles for the separation of the copolymers according to composition.

The second type is HPPLC (Table II). In HPPLC, the initial mobile phase is a nonsolvent for the sample copolymers, so that the sample copolymers injected into a column precipitate on the top of the column. Gradient elution is performed by adding a good solvent to the sample

Table II. Examples for HPPLC

Mobile Phase	Stationary Phase	Copolymer	Ref.
n-Hexane/THF	Silica gel Silica-RP-8	P(S–AN)	4
i-Octane/THF + MeOH	Silica-ODS	P(S–AN)	11
i-Octane/THF + MeOH	Silica gel	P(S–MMA)	12
n-Heptane/CH$_2$Cl$_2$ + MeOH	Silica gel	P(S–AN)	13
n-Heptane/CH$_2$Cl$_2$ + MeOH	Silica gel	P(S–MA), P(S–BA)	14

NOTE: S, polystyrene; AN, polyacrylonitrile; MMA, poly(methyl methacrylate); MA, poly(methyl acrylate); BA, poly(butyl acrylate).

copolymers. The copolymers that are precipitated on the top of the column start to redissolve into the mobile phase according to their composition with increasing content of the good solvent in the mobile phase (the selective dissolution according to their composition). Solubility of the copolymers in the mobile phase governs the separation of the copolymers according to composition, and the stationary phase does not have much effect on the separation. For example, 20% THF in n-hexane was the initial mobile phase and styrene–acrylonitrile copolymers injected into a column precipitated on the top of the column (4). By increasing the content of THF in the mobile phase, the copolymers started to redissolve in the order of increasing acrylonitrile content in the copolymers.

To classify a separation technique by LC into these two types, it should be clear whether the sample copolymers adsorb on the surface of the stationary phase or precipitate on the top of the column (phase separation) when the sample copolymers are injected into a column. If the separation mechanism is not clearly understood or when the separation by the solubility difference of the sample copolymers between the stationary and the mobile phases can be considered, then the technique can be classified into normal-phase and reversed-phase chromatography as the third type (Table III). Initial and final mobile phases should be good solvents for the sample copolymers. The initial mobile phases in Table III are 15% THF in acetonitrile (AcCN) or 10% THF in cyclohexane (15), 35% THF in n-hexane (16), 20% CH_2Cl_2 in AcCN (17), and 10% $CHCl_3$ in n-hexane (18). The final mobile phases are 65% THF in AcCN or 60% THF in cyclohexane (15), 85% THF in n-hexane (16), 80% CH_2Cl_2 in AcCN (17), and 40% $CHCl_3$ in n-hexane (18).

Stationary phases are hydrophobic or hydrophilic, and when the hydrophobic stationary phase such as silica-ODS is applied, the initial mobile phase is polar and the content of a less-polar solvent such as THF is increased in the mobile phase. This is called reversed-phase chromatography. In the case of a hydrophilic stationary phase (e.g., silica-CN), the initial mobile phase is less polar and the content of a

Table III. Normal- and Reversed-Phase Chromatography

Mobile Phase	Stationary Phase	Copolymer	Ref.
AcCN/THF	Silica-ODS, phenyl	P(S–MMA)	15
C_6H_{12}/THF	Silica-CN, NH_2	P(S–MMA)	15
n-Hexane/THF	Silica gel	P(S–MMA)	16
AcCN/CH_2Cl_2	PS gel	P(S–MMA)	17
n-Hexane/$CHCl_3$	PAN gel	P(S–PB)	18

NOTE: S, polystyrene; AN, polyacrylonitrile; MMA, poly(methyl methacrylate); PB, polybutadiene.

polar solvent to the initial mobile phase such as THF is increased in the mobile phase. This is called normal-phase chromatography. In both cases, the solubility of styrene-poor copolymers increases with increasing the content of the second solvent in the mobile phase and the styrene-poor copolymers start to elute from the column. This solubility difference of the copolymers between the stationary and the mobile phases is the main mechanism of the separation.

The fourth type, orthogonal chromatography, is a combination of SEC and NELC and uses polystyrene gel columns for SEC in both stages. THF is used as the mobile phase at the first stage and the sample copolymers are separated by size. Fractions separated by SEC are subjected to the second stage where the mobile phase is a mixture of THF and *n*-heptane (5). A mixture of THF and isopropanol was also used as the second mobile phase (*19*).

LAC at the critical point can be classified as the fifth type of NELC. Operating in the region between size exclusion and adsorption modes of LC by changing the composition of the bicomponent mobile phase, retention becomes independent of polymer size and the separation is accomplished exclusively by composition (*20*). This technique was applied to the characterization of block polymers (*21*).

Examples of LAC

Among several techniques for NELC, LAC with silica gel–chloroform (or DCE) + ethanol for the separation of styrene copolymers is easy to operate and has a wide applicability (*7*).

System and Samples.   Silica gel with a pore diameter of 3 nm and a mean particle size of 5 μm was packed in 4.6-mm i.d. $\times$ 50-mm-length stainless steel tubing. This column was thermostated at a specified temperature by using an air-oven or a column jacket. Gradient elution was performed as follows: the initial mobile phase (**A**) was a mixture of chloroform and ethanol (99.0:1.0, v/v), the composition of the final mobile phase (**B**) was 93.0:7.0 (v/v) chloroform–ethanol, and the composition of the mobile phase was changed from 100% (**A**) to 100% (**B**) in 30 min. The flow rate of the mobile phase was 0.5 mL/min (*8*). A UV detector was used at a wavelength of 260 nm (or 254 nm) or 233 nm (in the case of DCE) (*9*).

Samples tested were styrene copolymers of methacrylates, acrylates, vinyl acetate, and acrylonitrile, in addition to ethyl methacrylate–butyl methacrylate copolymers. These samples were dissolved in the initial mobile phase and the injection volume was 0.05–0.2 mL. These samples were prepared by solution polymerization at low conversion and have rather narrow CCD. These samples are random (statistical) copolymers.

Besides these random copolymers prepared by solution polymerization, those obained by bulk polymerization and block copolymers were also used for the application examples.

Methods of Separation. *Elution Behavior.* First, elution was performed by an isocratic elution mode. At a constant column temperature, the copolymers and homopolymers of polymethacrylates and polyacrylates were retained in the column with chloroform (and DCE) without ethanol. Only polystyrene could elute from the column. By adding ethanol to chloroform (and DCE), copolymers with a higher styrene content started to elute, and by increasing the ethanol content in the mobile phase, copolymers with less styrene were eluted.

For example, at a column temperature of 10 °C, poly(styrene–methyl methacrylate) P(S–MMA) copolymer with 48.7% styrene was retained in the column with chloroform–ethanol (99.5:0.5) and eluted 100% from the column with the mobile phase containing more than 1.5% ethanol (7) (Figure 1). At the mobile phase of chloroform–ethanol (99.0:1.0), half of the copolymer was retained in the column and the

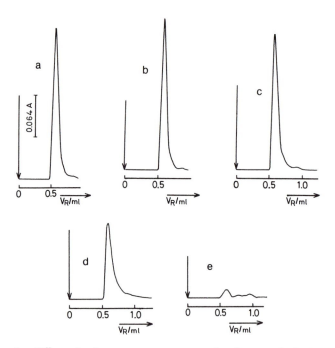

Figure 1.   Effect of column temperature on the elution of P(S–MMA) co-polymer having 66.3% styrene. Mobile phase, chloroform–ethanol (99.0: 1.0, v/v). Column temperature (°C), (a) 10, (b) 20, (c) 30, (d) 40, and (e) 50 (7).

rest was eluted from the column. The portion retained in the column had higher MMA (2.8%) than that eluted from the column (*1*). This result implies that even relatively homogeneous copolymers still have some chemical heterogeneity.

The copolymers tend to adsorb on the column at a higher column temperature, and copolymers with a higher methacrylate or acrylate component required a lower column temperature for elution. For example, a P(S–MMA) copolymer with 66.3% styrene eluted 100% from the column at column temperatures 10–30 °C with the mobile phase of chloroform–ethanol (99:1) and was retained in the column at 50 °C (Figure 2). The reason for the observation in Figure 2d was the same as that in Figure 1c. Lower column temperature (and/or a higher ethanol content in the mobile phase) was preferable for the elution of the copolymers having less styrene.

Mechanisms of Retention and Elution.   These results can be summarized as follows: the copolymers tend to adsorb in the column at a higher column temperature and at a lower content of ethanol in the mobile phase and the copolymers with a lower styrene component require a lower column temperature or a higher content of ethanol in the mobile phase to elute from the column. The ethanol content in the mobile phase or a column temperature did not affect peak retention volume for the copolymers. All the copolymers eluted at the same retention volume.

Carbonyl groups in the copolymers will hydrogen bond to silanol groups on the silica surface, and, consequently the copolymers will be adsorbed on the surface of silica gel. Neither chloroform nor DCE can displace the solutes from the surface. Ethanol is feasible to form hydrogen bonds to silanol groups and to control the content of the free silanol groups on the silica surface (*22*). Free silanol groups on the silica surface decrease in proportion to the ethanol content in the mobile phase. At elevated column temperature, ethanol that forms hydrogen bonds to silanol groups is desorbed and consequently free silanol groups on the silica surface increase. Lower ethanol content in the mobile phase and a higher column temperature result in the increase in free silanol groups on the silica surface and tend to have the copolymers adsorbed on the silica surface.

The population of carbonyl groups in the segment that contacts the surface of silica gel is in reverse proportion to the styrene content in the copolymers. Consequently, the copolymers with a smaller styrene content tend to adsorb on the surface of silica gel and tend to be retained in the column.

Gradient Elution.   To change the retention volume of the copolymers of different composition, a gradient elution mode was required

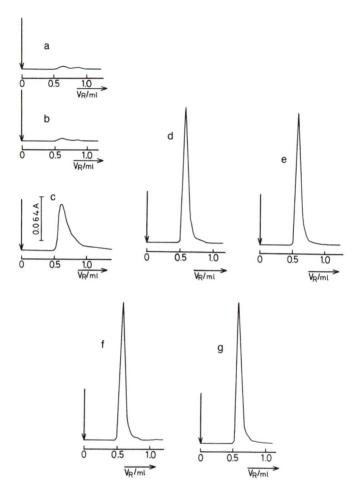

Figure 2. Effect of ethanol content in chloroform on the elution of P(S–MMA) copolymer having 48.7% styrene. Column temperature 10 °C. Mobile phase, chloroform–ethanol (a) 100:0, v/v, (b) 99.5:0.5, (c) 99.0:1.0, (d) 98.5:1.5, (e) 98.0:2.0, (f) 97.5:2.5, and (g) 97.0:3.0 (7).

at a specified column temperature (23). Under the gradient elution condition that the initial mobile phase was chloroform–ethanol (99:1), the final one was chloroform–ethanol (95.5:4.5), and the ethanol content was increased linearly in 15 min, P(S–MMA) copolymers with an MMA from 25% to 60% could be separated in the order of increasing MMA content at a column temperature of 80 °C and those with an MMA from 40% to 90%, at a column temperature of 30 °C.

Styrene copolymers of methyl, ethyl, and n-butyl acrylates and methacrylates were also separated according to their compositions (8) (Figure 3). Part of poly(styrene–ethyl methacrylate) P(S–EMA) copol-

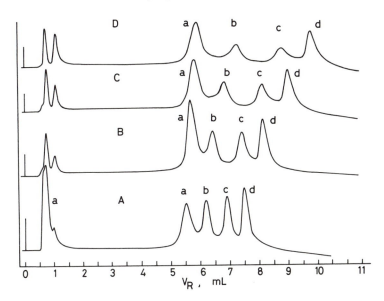

Figure 3.   LAC chromatograms of P(S–EMA) copolymers obtained by the linear gradient elution method. Column temperature (°C), (A) 40, (B) 50, (C) 60, and (D) 70. Sample, (a) P(S–EMA) I (styrene content 69.1%), (b) II (50.2%), (c) III (30.4), and (d) IV (15.5%). Gradient, chloroform–ethanol (99.0:1.0) to chloroform–ethanol (93.0:7.0) in 30 min (8).

ymer I eluted at the intersitial volume at a column temperature 40 °C. When the column temperature was increased to 50 °C, the four copolymers, P(S–EMA) I–IV, were retained in the column at first and then eluted at the appropriate retention volumes. Resolution was increased as column temperature and retention volume increased. Copolymers with the same styrene content that required much ethanol in the mobile phase to elute from the column at a constant column temperature were in the order of methyl, ethyl, and n-butyl acrylate (and methacrylate).

Plots of the relationship between the styrene content and retention volume for copolymers of styrene–acrylate and styrene–methacrylate with the same ester group lay roughly on the same line. This result indicates that a pair of copolymers with the same ester group and the same styrene content could not be separated (24). For example, copolymers of styrene–methyl acrylate and styrene–MMA with the same styrene content cannot be separated by this technique. In copolymers with the same styrene content, styrene–butyl acrylate and styrene–butyl methacrylate copolymers eluted first from a column, the copolymers of ethyl esters were next, and those of methyl esters eluted last.

SEC–LAC. To determine the molecular weight dependence of retention volume, the copolymers were fractionated by SEC into six fractions and LAC chromatograms were obtained. The peak retention volume of each fraction was almost the same. This result signified negligible molecular weight dependence (25). P(S–MMA) of narrow chemical composition distributions still had the compositional difference of several percent (23). For example, P(S–MMA) copolymer with 66.3% styrene was divided into two fractions by LAC and the composition of each fractions was measured. The difference in the composition was 2% (23).

Fractionation by SEC followed by LAC gave both MWD and CCD. P(S–MMA) with a broad CCD prepared by bulk polymerization was separated by SEC, and each fraction was then separated by LAC (26). An example of the MMA content range in a copolymer with a broad CCD (average composition, MMA 67.2%) was between 54% and 85%. Fractions having larger molecular weights had larger MMA content.

Application to Other Styrene Copolymers. The method presented here can be applied to other styrene copolymers that are feasible for hydrogen bonding to the silica surface. Styrene–MMA block copolymers of AB type were separated according to their composition by LAC (27). These block copolymer samples did not include random copolymers theoretically. Styrene–vinyl acetate block copolymers were characterized by both LAC and SEC (28). The copolymers were separated by LAC according to composition in order of increasing vinyl acetate content. There were positive trends in the composition and the molecular weight that the copolymer fractions containing more styrene have lower molecular weights.

Styrene–acrylonitrile copolymers also can be separated according to composition by this technique. Styrene-rich copolymers eluted first (29).

The Use of DEC. *Detection of Methacrylate (and Acrylate) Component.* DCE was transparent at wavelengths over 230 nm, and methacrylate (and acrylate) homopolymers and copolymers could be monitored with a UV detector around 233 nm (9). The molar absorption coefficients for both polystyrene and poly(methyl methacrylate) (PMMA) at 233 nm were nearly equal, and the chromatograms obtained at this wavelength reflected the relative amounts of the copolymers with different chemical compositions.

P(S–MMA) copolymers of the whole range of composition and PMMA were separated by a gradient elution method at a column temperature of 50 °C (9). The initial mobile phase was a mixture of DCE and ethanol (99:1, v/v), and the ethanol content was increased to 5.0% in 20 min and then to 10.0% in 5 min.

Ethyl methacrylate and *n*-butyl methacrylate homopolymers and copolymers were separated at a column temperature of 60 °C by gradient elution from a mixture of DCE–ethanol (99:1) to DCE–ethanol (90:10) in 20 min (*10*) (Figure 4).

Direct Characterization of CCD. As the chromatograms obtained at a wavelength of 233 nm represent the concentration of the copolymers independent of their composition (*9*), direct characterization of a CCD from the chromatograms monitored at 233 nm was possible with minor modification (*30*). When a calibration curve of retention volume versus styrene content is a straight line, the axis of abscissa (retention volume) can be converted directly to the scale of the copolymer composition. The calibration curve is normally not a straight line, and the chromatogram monitored at 233 nm should be converted by the following procedure. Normalize the chromatogram monitored at 233 nm. Divide the chromatogram into equal parts and measure the height of the chromatogram (dW/dV_R) at each divided point i. Calculate a slope of the calibration curve [$dV_R/d(\text{styrene \%})$] of retention volume versus styrene content at each divided point i. The ordinate for a CCD can be obtained by

$$\frac{dW}{d(\text{styrene \%})} = \frac{dW}{dV_R} \cdot \frac{dV_R}{d(\text{styrene \%})}$$

where W is the weight fraction of the copolymer. The abscissa for a CCD is converted from the scale of retention volume to styrene % by using the calibration curve.

Summary
The combination of SEC and NELC makes it possible to simultaneously characterize both an MWD and a CCD for several copolymers. NELC

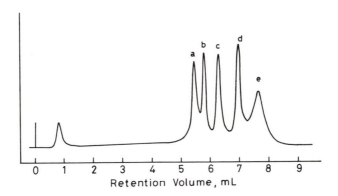

Figure 4. LAC chromatograms of ethyl methacrylate (EMA) and butyl methacrylate (BMA) hompolymers and copolymers. Column temperature, 60 °C. Sample, (a) PBMA, (b) EMA–BMA (25:75) copolymer, (c) EMA–BMA (50:50), (d) EMA–BMA (75:25), and (e) PEMA (10).

techniques can be classified into five types, and most styrene copolymers can be converted by these five techniques. The use of LAC with a system of silica gel–chloroform (or DEC) + ethanol for the separation of styrene copolymers was stressed.

References

1. Mori, S. *J. Chromatogra.* 1987, *411*, 355–362.
2. Teramachi, S.; Hasegawa, A.; Shima, Y.; Akatsuka, M.; Nakayama, M. *Macromolecules* 1979, *12*, 992–996.
3. Danielewicz, M.; Kubin, M. *J. Appl. Polym. Sci.* 1981, *26*, 951–956.
4. Glöckner, G.; Kroschwitz, H.; Meissner, C. H. *Acta Polymerica* 1982, *33*, 614–616.
5. Balke, S. T.; Patel, R. D. In *Polymer Characterization;* Craver, C. D., Ed.; Advances in Chemistry 203; American Chemical Society: Washington, DC, 1983; pp 281–310.
6. Mourey, T. H. *J. Chromatogr.* 1986, *357*, 101–106.
7. Mori, S.; Uno, Y. *Anal. Chem.* 1987, *59*, 90–94.
8. Mori, S.; Mouri, M. *Anal. Chem.* 1989, *61*, 2171–2175.
9. Mori, S. *J. Chromatogr.* 1991, *541*, 375–382.
10. Mori, S. *Anal. Chem.* 1990, *62*, 1902–1904.
11. Glöckner, G.; van den Berg, J. H. M.; Meijerink, N. L. J.; Scholte, T. G.; Koningsveld, R. *Macromolecules* 1984, *17*, 962–967.
12. Glöckner, G.; van den Berg, J. H. M. *J. Chromatogr.* 1986, *352*, 511–522.
13. Schultz, R.; Engelhardt, H. *Chromatographia* 1990, *29*, 325–332.
14. Sparidans, R. W.; Claessens, H. A.; Van Doremaele, G. H. J.; Van Herk, A. M. *J. Chromatogr.* 1990, *508*, 319–331.
15. Teramachi, S.; Hasegawa, A.; Motoyama, K. *Polym. J.* 1990, *22*, 480–496.
16. Sato, H.; Takeuchi, H.; Tanaka, Y. *Macromolecules* 1986, *19*, 2613–2617.
17. Sato, H.; Mitsutani, K.; Shimizu, I.; Tanaka, Y. *J. Chromatogr.* 1988, *447*, 387–391.
18. Sato, H.; Takeuchi, H.; Suzuki, S.; Tanaka, Y. *Makromol. Chem. Rapid Commun.* 1984, *5*, 719–722.
19. Dawkins, J. V.; Montenegro, A. M. C. *Br. Polym. J.* 1989, *21*, 31–36.
20. Schulz, G.; Much, H.; Krueger, H.; Wehrstedt, C. *J. Liq. Chromatogr.* 1990, *13*, 1745–1763.
21. Pasch, H.; Brinkmann, C.; Much, H.; Just, V. *J. Chromatogr.* 1992, *623*, 315–322.
22. Mori, S. *J. Liq. Chromatogr.* 1989, *12*, 323–336.
23. Mori, S.; Uno, Y. *J. Appl. Polym. Sci.* 1987, *34*, 2689–2699.
24. Mori, S. *J. Appl. Polym. Sci.: Appl. Polym. Symp.* 1990, *45*, 71–85.
25. Mori, S. *Anal. Sci.* 1988, *4*, 365–369.
26. Mori, S. *Anal. Chem.* 1988, *60*, 1125–1128.
27. Mori, S. *J. Appl. Polym. Sci.* 1989, *38*, 95–103.
28. Mori, S. *J. Chromatogr.* 1990, *503*, 411–419.
29. Morita, A. M.S. Thesis, Mie Univesity, Tsu, Mie, Japan, 1989.
30. Mori, S. *J. Appl. Polym. Sci.: Appl. Polym. Symp.* 1991, *48*, 133–139.

RECEIVED for review January 6, 1994. ACCEPTED revised manuscript June 17, 1994.

Two-Dimensional Chromatography for the Deformulation of Complex Copolymers

P. Kilz,[1] R.-P. Krüger,[2] H. Much,[2] and G. Schulz[2]

[1] Polymer Standards Service GmbH, P.O. Box 3368, D–55023 Mainz, Germany
[2] Center for Macromolecular Chemistry, Rudower Chaussee 5, D–12489 Berlin, Germany

A totally automated characterization of complex copolymers and blends by two-dimensional (2D) liquid chromatography–size-exclusion chromatography (LC–SEC) is described. The analysis of a 16-component mixture of a star block copolymer with wide range of molar masses and chemical composition by gradient high-performance LC or SEC alone does not give correct information. An on-line combination of both methods leads to much improved resolution. All 16 components are separated in the 2D contour map, enabling accurate and easy integration. The 2D LC–SEC analysis of polyester samples revealed the existence of several end groups, previously undiscovered. These byproducts influence mechanical and physical-chemical properties of polyesters. 2D chromatography allows the comprehensive and reliable separation and deformulation of complex analytes with very high resolution. Depending on the task, different types of separation methods may be combined to give best results. 2D chromatography is a powerful tool that may be used to understand and correlate macroscopic behavior of complex polymers to the molecular level.

THE CHARACTERIZATION OF MODERN HIGH-PERFORMANCE POLYMERS is still a challenge for polymer scientists. Copolymers and complex polymer blends play an important role in many applications (1). A fast, reliable, and comprehensive method is needed to succeed in this task. Size-exclusion chromatography (SEC) is a standard method for the determination of molar mass distributions (MMDs) and molecular weights, if proper calibration has been performed. Since its early days, efforts have been made to use SEC in the characterization of copolymers (2–4). Some approaches have proven successful in special cases, such as the characterization of block copolymers or copolymers with known and homogeneous composition (5–7). The introduction of reliable on-line viscometric detection has also helped to get more accurate information about copolymers (8–10). Because SEC separation is not based on molar mass but on molecular size, analysis of complex samples like copolymers and polymer blends modified by additives, plasticizers, and various stabilizers by SEC alone is generally not sufficient.

A combination of several separation methods or versatile detection has been used with great success in other analytical areas to characterize complex analytes [e.g., gas chromatography (GC) coupled to mass spectrometry, on-line coupling of liquid chromatography to gas chromatography (LC–GC)]. Two-dimensional (2D) chromatography of polymers (also known as cross-fractionation or orthogonal chromatography) has been used only infrequently in the past (11, 12) due to huge instrumental requirements and the difficulty of data reduction and presentation.

The versatility of the 2D approach is illustrated with a four-arm star-shaped block copolymer (a mixture of 16 components), which was synthesized in our laboratories to understand and demonstrate the advantages of 2D chromatography. These 16 components are a mixture of four different styrene–butadiene (St–Bd) copolymer compositions, each consisting of four molar masses (the St–Bd precursor with one to four arms). Another polymer investigated is a real-world application: the deformulation of a telomeric aliphatic polyester.

Theory

The main feature of polymers is their MMD, which is well known and understood today. However, several other properties in which the breadth of distribution are important and influence polymer behavior (see Figure 1) include physical, the classical chain-length distribution; chemical, two or more comonomers are incorporated in different fractions; topological, polymer architecture may differ (e.g., linear, branched, grafted, cyclic, star or comb-like, and dendritic); structural, comonomer placement may be random, block, alternating, and so on; and functional, distribution of chain functions (e.g., all chain ends or only some carry specific groups). Other properties the polymers may disperse (tacticity and crystallite dimensions) are not of the same general interest or cannot be characterized by solution methods.

The main disadvantage of SEC is its inability to quantitatively distinguish different polymer architectures and chemical heterogeneity. The SEC separation is governed by molar size, which is influenced by

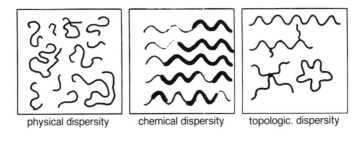

physical dispersity   chemical dispersity   topologic. dispersity

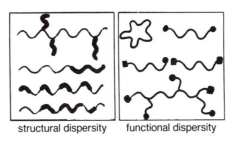

structural dispersity   functional dispersity

Figure 1. Various property distributions in polymers that are accessible by chromatography.

chain length, chemical composition, polymer topology, and so on. Several methods have been proposed and used to obtain chemical composition and molar mass information in the same chromatographic run:

1. Multiple detection SEC systems: n independent detector signals (different responses by components) allow the composition calculation of n components in the sample (copolymer or blend).

2. Universal calibration: measurement of Mark–Houwink coefficients for copolymers with homogeneous and known composition will give copolymer molar masses.

3. SEC with viscometric detection: this elegant method permits the on-line measurement of Mark–Houwink coefficients for copolymers of various architectures; copolymer M_n measurement applying Goldwasser's formula (13) is an additional benefit.

4. SEC with light-scattering detection: direct copolymer molar mass measurement for chemically homogeneous and segmented copolymers independent of their structure.

Table I shows these different approaches and their requirements, benefits, and limitations of chromatographic copolymer analysis.

Table I. Chromatographic Methods for Copolymer/Blend Characterization

Method	Requirements	Preconditions	Advantages	Limitations	Refs.
Multiple detection	2+ detectors, proper calibrants	No segment–segment interactions, No neighbor-group effects	Bulk composition and composition distribution, Broad applicability, Intrinsic generation of calibration curve, No additional sample preparation work	Statistical copolymers, Densely grafted chains	2, 3, 5, 8
Universal calibration	Base calibration, $[\eta]$-M relationship	Universal calibration, Homogeneous composition	Simple, Accurate MMD	Full knowledge about samples, Chemically homogeneous samples only	5–7, 10
Light-scattering detection	Light-scattering detector	Known dn/dc as a function of elution	Direct MMD measurement, No calibration, Architecture independence	No CCD information, No r_gs	5, 8
Viscometric detection	Viscometric detector	Universal calibration	Easy MMD calculation, K, for copolymers, Architecture independence	No CCD information, No heterogeneous samples	8, 9 7, 10, 13
2D chromatography	Two chromatographs (special software)	Methods for each dimension	This work	This work	11, 12, 14, this work

NOTE: CCD, chemical composition distribution.

The introduction of molar mass-sensitive detectors overcame problems in SEC analysis of various polymer topologies, if the chromatographic technique is able to separate them properly (*8*).

Gradient high-performance liquid chromatography (HPLC) has been useful for the characterization of copolymers (*14–19*). In such experiments, careful choice of separation conditions is a *conditio sine qua non.* Otherwise, low resolution for the polymeric sample will obstruct the separation. However, the separation in HPLC, dominated by enthalpic interactions, perfectly complements the entropic nature of the SEC retention mechanism in the characterization of complex polymer formulations.

However, HPLC sorbents also show SEC behavior to some extent dependent on the pore size of the stationary phase relative to the molar size of the solute. Copolymers with the same composition but different molar masses will in general have somewhat different retention characteristics. This may lead to copolymer HPLC fractions with heterogeneous chemical compositions and may contain some chains with different molar mass and comonomer content.

Entelis et al. (*20*) found that homopolymers of different molar masses show exactly similar retention behavior on silica if a special eluent mixture was used. They found that under "critical" conditions the sorbent did not "see" the polymeric nature of the chain. The separation was dependant only on the enthalpic interaction of the sample–sorbent pair [so-called "critical chromatography" or "liquid adsorption chromatography under critical conditions" (LACCC)] (*20–24*).

We tried to combine the advantages of HPLC and SEC by using a fully automated, software controlled, 2D chromatography system for the on-line analysis of composition, end-group functionality, and MMD. It consists of two chromatographs, one that separates by chemical composition (e.g., a gradient HPLC) and a SEC instrument for subsequent separation by size (cf. Figure 2).

2D Approach

Based on previous off-line results by other authors (*11, 12, 14*), we anticipated improved resolution of multicomponent analytes. We also hoped for chemically homogeneous fractions to measure MMDs of copolymers and blends correctly.

In contrast to previous publications, we wanted to create chemically homogeneous fractions in the first separation and work on these for molar mass characterization. Therefore, HPLC (or LACCC) was the first and SEC the second separation method. The advantages of this setup are numerous. LC has more parameters (gradient, stationary phase, etc.) to adjust the separation according to the chemical nature of the

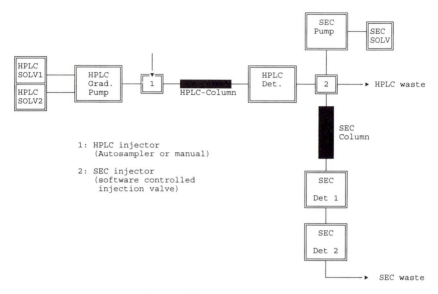

Figure 2.   *Experimental setup of the 2D chromatography system (horizontal, LC components; vertical, SEC components).*

sample and has better fine-tuning (gradient) elution (giving "cleaner" fractions). Also, the higher sample load on LC than on SEC columns gives a higher signal-to-noise ratio in the second dimension. Therefore, the first dimension (LC) will separate by the chemical nature of the sample, and the subsequent SEC analysis of each fraction will measure the molar mass dependence for each chemical composition. The chemical composition of all eluting species can be calculated by detector response calibration or by model compounds separately in each dimension.

This can be done, for example, by multiple detection, where the absolute comonomer concentration, w_k, is measured for each analytical fraction after concentration detector calibration according to

$$U_d(V) = \sum_k f_{dk} \cdot w_k(V) \tag{1}$$

The knowledge of the response factors f_{dk} for each detector, d, and component, k, allows for the absolute measurement of the comonomer concentration (w_k) in all detector cells (d) at all elution volumes.

This approach is basically universal and requires only n pure reference samples for n comonomers in copolymers or components in polymer blends. However, neighbor-group effects have to be absent, which might otherwise influence composition calculations for copolymers.

Additionally or alternatively, composition calibration in the LC dimension is possible, using pure reference material or model compounds depending on the availability of such samples. The advantage of this method is its simplicity and independence of any preconditions or assumptions. However, it requires a whole set of copolymer model compounds and highly reproducible LC elution (gradient formation).

When the chemical composition at each slice is known (and each fraction is chemically pure) the molar mass of the copolymer, M_c, can be calculated easily by Runyon's empirical approach (2):

$$\lg M_c(V) = \sum_k w_k(V) \cdot \lg M_k(V) \tag{2}$$

where $M_k(V)$ represents the calibration curve for each homopolymer of comonomer k.

In this kind of calculation the absence of segment–segment interactions and chemically monodisperse SEC fractions has to be assumed. The main benefits of this approach is ordinary SEC equipment is used and the copolymer analysis is done with the same injection without additional sample preparation.

Experimental Procedures

Chemicals. The four-armed star St–Bd block copolymers with different Bd compositions were synthesized in our laboratories at PSS (Mainz, Germany) by anionic polymerization according to standard procedures (25, 26) modified to give samples with well-known structure and molar mass control.

Synthesis of Model Polymers by Anionic Polymerization. All coproducts have the same chemical composition (St–Bd ratio) [A, styrene; B, butadiene-1,3; linear (M) + ext. linear ($2M$) + three-arm star ($3M$) + four-arm star ($4M$)]:

$$I^\ominus + mA \longrightarrow I\text{-}A_m^{\ominus} \xrightarrow{+ nB} I\text{-}A_m\text{-}B_n^{\ominus} \xrightarrow[\text{part. term.}]{+ XR_4} I\text{-}A_m\text{-}B_n$$

$$+ \ I\text{-}A_m\text{-}B_n\text{-}B_n\text{-}A_m\text{-}I \ + \ \begin{matrix} I\text{-}A_m\text{-}B_n \\ \diagdown \\ X \\ \diagup \\ I\text{-}A_m\text{-}B_n \end{matrix} \begin{matrix} B_n A_m\text{-}I \\ \diagup \\ \\ \diagdown \\ B_n\text{-}A_m\text{-}I \end{matrix} \ + \ \begin{matrix} I\text{-}A_m\text{-}B_n \\ \diagdown \\ X \\ \diagup \\ I\text{-}A_m\text{-}B_n \end{matrix} \begin{matrix} B_n\text{-}A_m\text{-}I \\ \diagup \\ \\ \diagdown \\ B_n\text{-}A_m\text{-}I \end{matrix}$$

The preparations were carried out in a way to give all structures that are theoretically possible: St–Bd precursor (of molar mass M_{arm}), linear (St–Bd)$_2$ ($M = 2M_{arm}$), the three-arm product (St–Bd)$_3$ ($M = 3M_{arm}$), and, finally, the four-arm compound (St–Bd)$_4$ ($M = 4M_{arm}$). Four samples with varying Bd content (20, 40, 60, and 80%) were prepared in this way. A mixture of these samples was used to check the capabilities of the 2D chromatography system.

The sample mixture can be represented by a (4 × 4) matrix with Bd composition (X_i) and molar mass (M_i) as variables:

$$\begin{pmatrix} C_{M_1,X20} & C_{M_2,X20} & C_{M_3,X20} & C_{M_4,X20} \\ C_{M_1,X40} & C_{M_2,X40} & C_{M_3,X40} & C_{M_4,X40} \\ C_{M_1,X60} & C_{M_2,X60} & C_{M_3,X60} & C_{M_4,X60} \\ C_{M_1,X80} & C_{M_2,X80} & C_{M_3,X80} & C_{M_4,X80} \end{pmatrix}$$

where $C_{M_1,X20}$ represents the concentration C of species with molar mass M_1 and 20% butadiene content.

The telomeric aliphatic polyesters were produced by polycondensation based on adipic acid and hexamethylene glycol in various stoichiometric amounts to generate polyesters of different end group functionality. The polyesters of different molar mass and corresponding reference samples were synthesized at the Center for Macromolecular Chemistry, Berlin, Germany. These types of polyesters are widely used as lacquers and precursors for the production of several important polyurethanes.

Linear polystyrene and polybutadiene-1,4 standards (PSS) were used for the calibration of the SEC columns. The columns for the polyester samples were calibrated by resolved oligomeric peaks and by matching model compounds.

Gradient-HPLC System. A Spectra Physics 8840 (Munich, Germany) pump was used for gradient formation and solvent (i-octane–tetrahydrofuran (THF), linear gradient, 20–100% THF) delivery. The samples were injected by a Rheodyne 7125 manual injection valve. A nonmodified SiO_2 column with 60 Å porosity (300 × 8 mm dimension) was used for the separation in the first dimension. A Spectra Physics 8450 UV–VIS detector monitored the concentration profile. Transfer to the second dimension (SEC instrument) was performed by an electrically actuated Rheodyne 7010 valve equipped with a 100-μL sample loop. The injector was kept in load position until the sample to be injected into the SEC dimension was inside the loop. Injection was software controlled by timed events and contact closure.

LACCC System. The first dimension in the 2D polyester characterization consisted of a Jasco 880PU pump delivering the acetone–hexane (50.5%:49.5%) solvent mixture, which are the critical conditions for the separation of the polyester samples using one 7-μm Si-120 silica column (Tessek, Prague, Czechoslovakia) for the LACCC separation. A Erma 7511 refractive index (RI) detector was used to monitor the concentration trace of the fractions.

SEC System. A Spectra Physics IsoChrom pump controlled the THF flow in the SEC part of the 2D instrument. Two SDV 5-μm SEC columns with 1000 and 10^5 Å porosity (PSS) were used for size separation of the HPLC fractions. UV (SP 8450, Spectro Physics) and RI (Shodex SE 61, Düsseldorf, Germany) detection allowed for conventional or multiple-detection data processing of detector traces. For the polyester analysis the SEC columns (50 and 100 Å) were operated in acetone as eluent.

All detectors and injectors were connected to a PSS GPC 2000 Data Station. Data acquisition, timed events, and all calculations were done with 2D-CHROM hardware and software (V 2.0) from PSS.

Results and Discussion

Multicomponent Star Block Copolymer on Gradient HPLC–SEC System. The 16-component star block copolymer mixture was injected into a gradient HPLC (*i*-octane–THF) on a silica column (60-Å pore size) to get a good separation by chemical composition. These fractions were transferred automatically by an electrically driven injection valve into the second dimension (SEC in THF; cf. Figure 2).

Figure 3 shows the result of a SEC analysis for the 16-component star block copolymer mixture (top trace). In each case four different peaks are visible. They correspond to the four molar masses of the sample that consists of chains with one to four arms, each arm having the same chain length and composition. Despite the high resolution, the chromatogram does not give any hint of the very complex nature of this sample. Even when pure fractions with different chemical composition were run in the SEC instrument (traces 2–4 in Figure 3), the retention behavior does not show significant changes as compared with the sample mixture. In each case a tetramodal MMD is visible, indicating the copolymers of different polymer architecture and molar mass. The SEC

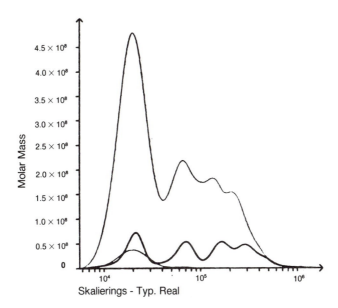

Figure 3. Comparison of MMD for composition mixture of star block copolymers with samples of set Bd content; no significant change in SEC behavior is visible with varying Bd content. Trace 1, MMD of sample mixture; trace 2, SEC separation of individual sample with 80% Bd content; trace 3, SEC result of 60% Bd sample; trace 4, MMD of copolymer with 20% Bd content.

separation alone does not show any difference in chemical composition of the samples, which varied from 20 to 80% Bd content.

Running the same sample mixture in gradient HPLC alone gives poorly resolved peaks, which may suggest different composition but not of different molar masses and structures (cf. Figure 4).

The combination of the two methods in the 2D setup dramatically increases the resolution of the separation system and gives a clear picture of the complex nature of the sample mixture. A three-dimensional (3D) representation of the gradient HPLC–SEC separation is given in Figure 5; each trace represents a fraction transferred from HPLC into the SEC system and gives the result of the SEC analysis. The 3D view already indicates the complexity of the sample mixture. The point of view can be chosen deliberately in the software. Based on the 2D analysis, a contour map with 16 spots would be expected. Each spot would represent a component within the complex sample that is defined by a single composition and molar mass. The contour map should also reflect the (4 × 4)

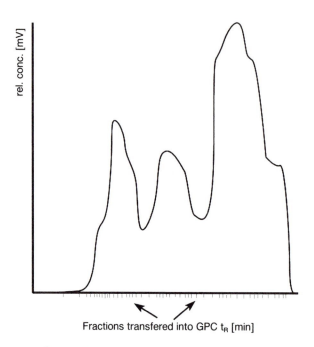

Figure 4. Gradient HPLC separation of a 16-component star-shaped St–Bd copolymer with four different molar masses, each having four different compositions also.

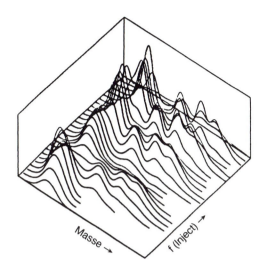

Figure 5.   3D plot of the HPLC–SEC analysis of the complex star-shaped St–Bd block copolymer.

concentration matrix of the sample (cf. previous example). The result of the theoretical 2D separation is shown as a contour map in Figure 6. The experimental evidence of a very much improved resolution in the 2D analyses is shown in Plate 1. This contour map is calculated from experimental data based on 28 transfer injections from the gradient HPLC into the SEC part of the 2D system.

The contour plot clearly reveals the chemical heterogeneity (y-axis, chemical composition) and the MMD (x-axis) of the test mixture. The relative concentrations of the components are indicated by colors (a color reference chart is given on the right side). Contour maps can be read like topological maps; concentrations correspond to the altitude, first (second) dimension separation corresponds to the north (east) direction. Sixteen major peaks are resolved with high selectivity. These correspond directly to the components in the sample mixture. Some byproducts are also revealed, which were not detected in LC or SEC alone. A slight molar mass dependence of the HPLC separation is visible. This kind of behavior is normal for polymers run on HPLC phases, because pores in the HPLC sorbents lead to size-exclusion effects that overlap with polymer–sorbent surface interaction. Consequently, 2D separations will in general be not orthogonal but skewed depending on the sorbent pore size and its distribution. Such kinds of observations are also in accordance with experiments by other authors (*14*) and corroborates the high performance of the separation system and the accuracy of the PSS 2D-CHROM software. Reproducible integration and full quantification of peaks in the contour map are available in the 2D-

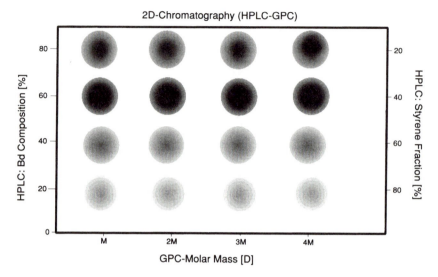

Figure 6. Theoretical contour map of the 2D HPLC–SEC analysis of the 16-component mixture.

CHROM software. It is also possible to use molar mass sensitive detectors in this setup to get independent of calibrations or assumptions for the calculation of molar masses for copolymers with unknown structures. Multidetection analysis for independent copolymer composition calculations (8) and 2D segment distributions are already fully implemented in the PSS 2D-CHROM software package.

End-Group Analysis of Polyester on LACCC–SEC System. Chemically homogeneous polymers with broad MMD show a narrow peak at critical conditions. In the case of segmented copolymers, the molar mass dependence of one component is absent, allowing the polydispersity of another segment in graft, block, or comb-shaped copolymers to be studied. Different end groups may even be investigated by this technique if chain lengths are small enough. In this part of our work critical chromatography (LACCC) was used to try to correlate the performance of polyesters to different end groups. Regular end groups in the condensation polymerization of polyesters may be diol and diacid functions or mixed OH and COOH end groups. The kind of end groups and their relative concentrations can be controlled by the molar ratios of monomeric diacid and monomeric glycol. Additional treatment of the polyester (e.g., thermal postcondensation) may change the type of end group and influence the mechanical and physicochemical properties.

Figure 7b shows the high-resolution SEC chromatogram of polyesters of different molar mass. The high efficiency of the columns permits

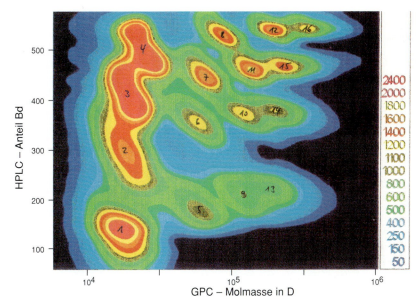

Plate 1. *Experimental contour map of the 16-component star-shaped St–Bd block copolymer in gradient HPLC–SEC separation. Fractionation by composition (HPLC) on ordinate; size separation (SEC) on abscissa; all 16 species can be isolated in very good yield.*

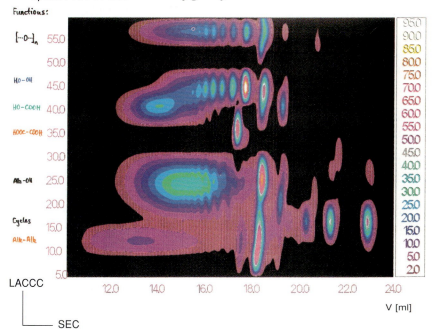

Plate 2. *Contour map of polyester with an acid number of 0.2 in a 2D LACCC–SEC separation, end-group dependence as detected by LACCC is plotted on y-axis (examples of terminal functions are also given); SEC elution volume on x-axis.*

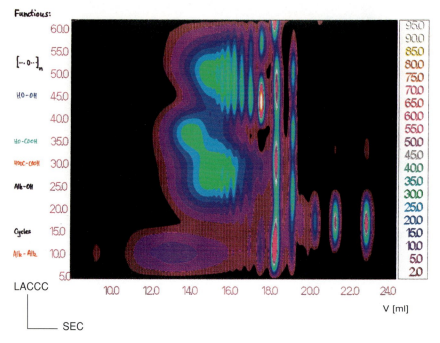

Functions:

[...O...]ₙ

HO-OH

HO-COOH

HOOC-COOH

Alk-OH

Cycles

Alk-Alk

LACCC

SEC

V [ml]

Plate 3.   Contour map of polyester with acidic end groups (acid number 5.0) in a 2D deformulation experiment.

the separation and identification of discrete oligomers. All oligomers show very narrow band width and no shoulders, which might be interpreted as an absence of byproducts. Oligomers from samples prepared under different condensation conditions show exactly the same retention behavior using St–divinyl benzene columns and acetone as mobile phase. These observations might lead to the conclusion that all of the oligomers were chemically identical.

However, a gradient HPLC separation of the same sample exhibits a very complex pattern (cf. Figure 7d). This is a clear indication of the existence of many more byproducts than were even anticipated from theory. Identification and quantification of these side peaks is very difficult, because separation efficiency for fractions with higher molar mass dramatically decreases. This complex pattern becomes much simpler if a LACCC separation is performed as shown in Figure 7c. A main peak surrounded by satellite peaks of different end-group functionality is found. Identification and quantification become much much easier due to the simple peak pattern and superb resolution. In our case, all peaks were identified by model compounds possessing the appropriate end groups. The assigned terminal functions are annotated in Figure 7c. Figures 7b–7d, correspond to different modes of chromatography as shown in the corresponding schematic in Figure 7a.

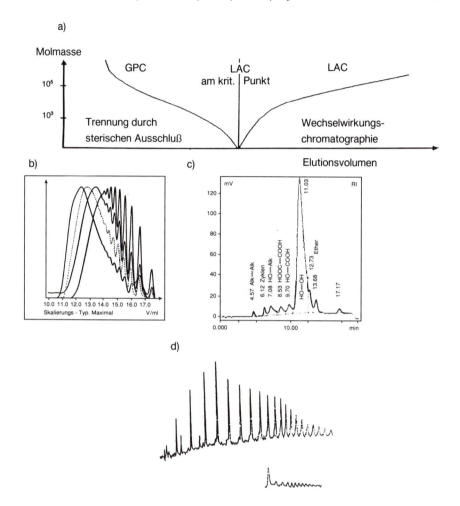

Figure 7.   *Examples of different modes of chromatography: (a) calibration curves (schematic), (b) SEC separation of oligomeric polyesters with different molar mass, (c) critical chromatogram (LACCC) of polyester; diole terminated sample is main compound, and (d) gradient HPLC run of the same polyester sample showing multiple distributions.*

Size-exclusion behavior is governed by entropy changes of the solute between mobile and stationary phase. Figure 7b is an example of polyester analysis under SEC conditions. When the polarity of the eluent is reduced, adsorption is the dominant influence in the separation. The solute is retained longer in the column due to enthalpic interactions with the sorbent. An example of an LC polyester separation using the same column but a less polar eluent mixture is given in Figure 7d. SEC and LC can be considered to be the two extremes of chromatographic behavior. There has to be some technique with balanced enthalpy and entropy where there is no change in the free energy during elution. This is a critical state; separations under such conditions are called critical chromatography (or LACCC). In this separation mode all homogeneous samples elute at the same retention volume despite their molar mass. An example for polyester analysis using LACCC technique is given in Figure 7c. The optimization of eluent composition for LACCC separations is shown in Figure 8. Figure 8a shows the retention behavior (on the abscissa) of samples of increasing molar mass (plotted in relative scale on the ordinate) in the three different modes of chromatography. Derived calibration plots (log M_w vs. V_e) from these chromatograms are given in Figure 8b; eluent composition for each separation mode is also shown.

To study the influence of polymer structure, molar mass, and end groups on the performance of these polyesters, 2D LACCC–SEC separations were carried out. The contour map that is most useful for quantitative analysis and interpretation is reproduced in Plate 2. The ordinate is proportional to the LACCC retention of the polyester. Specific end groups of polyester model compounds are also shown as a guideline. These model compounds with various end group and polymer structure characteristics were run under identical conditions

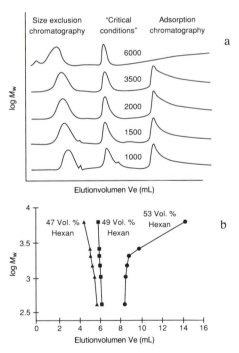

Figure 8. Optimization of eluent composition for the LACCC separation of polyester: (a) molar mass dependence of retention in different chromatographic separation modes; and (b) corresponding calibration curves.

in off-line LACCC experiments. The LACCC retention is plotted versus SEC elution volume to measure the molecular size of the LACCC fraction.

The contour plot reveals many important features of the sample. To the big surprise of many synthetic chemists working with polyesters, we could identify a broad range of molar masses and end-group functions in the sample. The 2D contour map reveals that each end group is formed in a specific molar mass range. With increasing LACCC retention, the following terminal functions can be identified (from bottom to top in the contour map; cf. Plate 2):

- **Alk–P_n–Alk** polyester end groups up to high molar masses with high concentration of some oligomers
- **Cycles –P_n–** of polyester chains (at high SEC elution volumes) in significant amounts
- **HOOC–P_n–OH** terminated chains over the whole molar mass range
- **HOOC–P_n–COOH** formed only as a single species
- **HOOC–P_n–OH** in predominantly long chains with lower oligomers missing
- **HO–P_n–OH** found mainly in oligomeric chains
- **Ether –O–** formed in significant quantities throughout all molar masses

A different adipic acid–hexane glycol polyester sample prepared with an excess of adipic acid shows characteristic differences in the 2D separation as it is easily revealed by its contour map (Plate 3). We observe the same end groups in similar molar mass ranges as before but in varying concentrations. The most obvious change can be seen in the center of the contour map, where there is a strong area in which the diacid terminated polyester chains elute. These chains have lower molar mass as compared with the regular polyester with HOOC–P_n–OH terminal groups. These results are in very good agreement with end-group titrations, which give an acid number of 5.0 for this sample as compared with an acid number of 0.2 for the polyester sample discussed in Plate 2. The 2D analysis of the previous sample showed no polymeric diacid functions at all.

Quantitative analysis of these 2D analyses shows pronounced differences for various end groups with molar mass and synthetic conditions (cf. Table II). Mechanical tests have shown that the mechanical stability of polyesters can be correlated to the content of cycles in the specimen under investigation.

Conclusions

Several methods can be used for the chromatographic characterization of complex polymers such as copolymers and polymer blends. The

Table II.   Quantitative Analysis of Various End Groups of Adipic Acid–
Hexane Glycol Polyesters as Determined by Critical Chromatography

	Molar Mass			
End Group	1000	1500	2000	3500
Alk–P_n–Alk	0.22	1.28	1.15	2.11
Cycles –P_n–	.054	1.64	1.81	3.22
HOOC–P_n–COOH	1.21	5.14	4.89	10.13
HOOC–P_n–OH	1.33	7.70	6.07	11.53
HO–P_n–OH	76.80	76.32	76.44	59.64
Ether –O–	16.95	11.84	10.25	9.57

NOTE: Values are functionality by weight percent.

application of a single method, such as multidetector SEC or SEC with light-scattering detection, for the analysis of such samples always has some limitations. The best way to overcome real or potential problems in the analysis of complex polymers or blends is 2D chromatography. Depending on the separation methods applied in the 2D chromatograph, fractionation according to various properties can be achieved.

The HPLC–SEC analysis of a 16-component star block St–Bd copolymer showed very good selectivity. The gradient HPLC could be optimized to fractionate by chemical composition; subsequent SEC automatically characterized the fractions for their molar mass dependence. All 16 components could be isolated and could be quantified.

The deformulation of a polyester was done with liquid-adsorption chromatography under critical conditions (LACCC) that was coupled to an SEC system. The dimension was separated by the end-group functionality to study byproducts formed during polycondensation. By 2D chromatography a number of species (e.g., cycles, ethers, alkyl terminated chains, and so on) could be identified. Some of them influenced the mechanical and thermal properties of the polyesters significantly.

The high resolution of LC–SEC separations and the full automation using 2D-CHROM software enable the reliable and comprehensive characterization and deformulation of complex analytes like copolymers, polymer blends, and additives. 2D chromatography will become a powerful tool with flexible and easy-to-use software. Basically, all types of LC methods can be combined to give superior resolution and reproducibility.

Identification and quantitation can be done with model compounds or with specific detection, like Fourier transform infrared or other kinds of multiple detection. However, 2D chromatography is also extremely useful for fingerprinting and visual comparison.

References

1. Fachtagung "Angewandte instrumentelle Analytik für Formmassen und Fertigteile aus polymeren Werkstoffen"; F. H. Würzburg-Schweinfurt: Würzburg, Germany, 1990.
2. Runyon, J. R.; Barnes, D. E.; Rudel, J. F.; Tung, L. H. *J. Appl. Polym. Sci.* **1969**, *13*, 2359.
3. Dondos, A.; Rempp, P.; Benoit, H. *Makromol. Chem.* **1969**, *130*, 223.
4. Ho-Duc, N.; Prud'homme, J. *Macromolecules* **1973**, *6*, 472.
5. Kilz, P.; Johann, C. *J. Appl. Polym. Sci. Symp.* **1991**, *48*, 111.
6. Chang, F. S. C. *J. Chromatogr.* **1971**, *55*, 67.
7. Goldwasser, J. M.; Rudin, A. *J. Liq. Chromatogr.* **1983**, *6*, 2433.
8. Kilz, P.; Gores, F. In *Chromatography of Polymers: Characterization by SEC and FFF*; Provder, T., Ed.; ACS Symposium Series 521; American Chemical Society: Washington, DC, 1993; Chapter 10.
9. Yau, W. W. *Chemtracts, Macromol. Chem.* **1990**, *1*, 1.
10. Goldwasser, J. M. In *Chromatography of Polymers: Characterization by SEC and FFF*; Provder, T., Ed.; ACS Symposium Series 521; American Chemical Society: Washington, DC, 1993; Chapter 16.
11. Balke, S. T.; Patel, R. D. *J. Polym. Sci., B. Polym. Lett. Ed.* **1980**, *18*, 453.
12. Ogawa, T.; Sakai, M. *J. Polym. Sci., Polym. Phys. Ed.* **1982**, *19*, 1377.
13. Goldwasser, J. M. *Proceedings of the International GPC Symposium*; **1989**; p 150.
14. Glöckner, G. *Gradient HPLC of Copolymers and Chromatographic Cross-Fractionation*; Springer: Berlin, Germany, 1991.
15. Glöckner, G.; Koschwitz, H.; Meissner, C. *Acta Polym.* **1982**, *33*, 614.
16. Sato, H.; Takeuchi, H.; Tanaka, Y. *Macromolecules* **1986**, *19*, 2613.
17. Danielewicz, M.; Kubin, M. *J. Appl. Polym. Sci.* **1981**, *26*, 951.
18. Mori, S. *J. Appl. Polym. Sci.* **1989**, *38*, 95.
19. Mourey, T. H. *J. Chromatogr.* **1986**, *357*, 101.
20. Entelis, S. G.; Evreinov, V. V.; Gorshkov, A. V. *Adv. Polym. Sci.* **1986**, *76*, 129.
21. Belenkii, B. G.; Gankina, E. S. *J. Chromatogr.* **1977**, *141*, 13.
22. Schulz, G.; Much, H.; Krüger, H.; Wehrsted, G. *J. Liq. Chromatogr.* **1990**, *13*, 1745.
23. Gorshkov, A. V.; Much, H.; Becker, H.; Pasch, H.; Evreinov, V. V.; Enteilis, S. G. *J. Chromatogr.* **1990**, *523*, 91.
24. Pasch, H.; Much, H.; Schulz, G.; Gorshkov, A. V. *LC GC Intl.* **1992**, *5*, 38.
25. Scwarc, M.; Levy, M.; Milkovich, R. *J. Am. Chem. Soc.* **1956**, *78*, 2656.
26. Corbin, N.; Prud'homme, J. *J. Polym. Sci., Polym. Chem. Ed.* **1976**, 1645.

RECEIVED for review January 6, 1994. ACCEPTED revised manuscript October 20, 1994.

18

Characterization of Block Copolymers Using Size-Exclusion Chromatography with Multiple Detectors

Elizabeth Meehan and Stephen O'Donohue

Polymer Laboratories Ltd., Essex Road, Church Stretton, Shropshire
SY6 6AX, United Kingdom

The characterization of block copolymers by size-exclusion chromatography (SEC) is complicated because the polymer may exhibit a chemical composition distribution (CCD) that will be superimposed on the molelcular weight distribution (MWD). To compensate for this difficulty, a dual detector approach using UV and differential refractive index (DRI) in series can be applied to give information regarding the CCD, although this method still relies on SEC column calibration to produce copolymer molecular weights. The addition of a third detector, low-angle laser light scattering, can be used to measure molecular weight directly without SEC column calibration, thus widening the applicability of SEC in the characterization of block copolymers. The characterization of two copolymer systems is described by using both the dual detector and the triple detector approaches.

SIZE-EXCLUSION CHROMATOGRAPHY (SEC) analysis, when applied to polymer characterization, requires the measurement of both concentration and molecular weight of each eluting species. For homopolymers, these measurements are readily achieved with a single concentration detector and a molecular weight calibration generated using well-characterized narrow or broad polydispersity polymer standards (*1*). For block copolymer systems, this standard approach is no longer applicable, because a chemical composition distribution (CCD), which must also be considered, may be superimposed upon the molecular weight distribution (MWD). This means for an eluting fraction of a given molecular size, the molecular weight and the detector response depends on the

0065–2393/95/0247–0243$12.00/0

composition at that elution volume. Therefore, the determination of MWD in block copolymer systems requires a knowledge of the CCD. The application of dual detection [UV and refractive index (RI)] to the SEC analysis of polystyrene–poly(methyl methacrylate) (PS–PMMA) has already been studied in this laboratory (2). Both MWD and CCD were determined using a methodology outlined by Runyon et al. (3). This approach relies on SEC column calibration with narrow polydispersity standards for each of the homopolymers as well as a measure of the detector response factors for each homopolymer to produce a copolymer MWD. In the case of PS and PMMA this is feasible, but in other block copolymer systems the availability of suitable molecular weight standards may be more limited. In addition, this procedure does rely on true SEC and is not valid for block copolymers for which the universal calibration does not hold true for both blocks in a given solvent system.

The work presented here is an alternative approach in which a third detector, a low-angle laser light-scattering detector (LALLS), is added to the system to give molecular weight information directly at each elution volume. This molecular weight information, combined with compositional information obtained from the two concentration detectors, UV and RI, yields accurate MWD and CCD for different block copolymer systems. Both methodologies outlined previously are applied to a PS–PMMA block copolymer for verification. The UV–LALLS–RI method is then applied to characterize a PS–polyethylene oxide (PS–PEO) block copolymer, which may not be characterized by using two concentration detectors alone.

Analytical Methods

The study involved three SEC detectors in series.

1. UV absorbance detector at 254 nm responding specifically to one of the homopolymers, PS. The response, Auv, is dependant on the instrument constant, Kuv, and the concentration, c, according to

$$Auv = Kuv \cdot c \qquad (1)$$

2. A LALLS detector from which the response Als can be used to measure molecular weight directly according to

$$Als = \frac{K' \cdot v^2 \cdot c \cdot M}{Kls} \qquad (2)$$

where K' is an optical constant

$$\frac{2\pi^2 n^2 (1 + \cos^2 \theta)}{\lambda^4 N_A}$$

n is the solvent refractive index, θ is the scattering angle, λ is the wavelength of the incident light, N_A is the Avogadro number, ν is the specific refractive index increment, M is the weight average molecular weight, and Kls is the instrument constant.

3. An RI detector where the response Ari is related to

$$Ari = Kri \cdot c \cdot \nu \qquad (3)$$

where Kri is the instrument constant.

Dual Detector Approach.   Using the RI and UV responses only, the calculation of copolymer molecular weight requires a knowledge of the weight fraction of each homopolymer and its molecular weight such that

$$\log M_{AB}i = W_A i \cdot \log M_A i + W_B i \cdot \log M_B i \qquad (4)$$

where $W_A i$ is the weight fraction of homopolymer A, $W_B i$ is the weight fraction of polymer B, $M_A i$ is the molecular weight of polymer A, and $M_B i$ is the molecular weight of polymer B at the same elution volume i. $W_A i$ and $W_B i$ can be measured via the area responses and the instrument constants for each homopolymer in each detector

$$W_A i = \frac{Ri \cdot K_B ri}{K_A uv - (K_A ri - K_B ri) \cdot Ri} \qquad (5)$$

where Ri is the ratio of the UV response to that of the RI response at the same elution volume i, $K_B ri$ is the RI instrument constant for polymer B, $K_A ri$ is the RI instrument constant for polymer A, and $K_A uv$ is the UV instrument constant for polymer A. $M_A i$ and $M_B i$ can be determined by conventional narrow standard SEC calibration.

These data treatments permit the calculation of the CCD and subsequently the MWD for a block copolymer system but still rely on SEC column calibration.

Triple Detector Approach.   In this case $M_{AB}i$ can be calculated directly from the RI and LALLS responses

$$M_{AB}i = \frac{Kri \cdot Kls \cdot Als}{K' \cdot Ari \cdot \nu} \tag{6}$$

The average ν value in equation 6 is calculated from the total response of the RI trace and the concentration according to equation 7

$$\nu = \frac{Ari}{Kri \cdot C} \tag{7}$$

Using the compositional information obtained from the dual detector UV–RI approach, the molecular weight of the copolymer obtained by this method may be further refined by correcting the $M_{AB}i$ at each slice with the νi distribution. The νi distribution may be calculated from

$$\nu_{AB}i = W_A i \cdot \nu_A + W_B i \cdot \nu_B \tag{8}$$

where $\nu_{AB}i$ is the copolymer specific RI increment at the elution volume i, ν_A is the specific RI increment of polymer A, and ν_B is the specific RI increment of polymer B.

Experimental Procedures

All SEC measurements were made using tetrahydrofuran (THF) (stabilized with 0.025% 2,6-di-*tert*-butyl-*p*-cresol, Fisons, Loughborough, UK) as eluent. The eluent was filtered four times through a 0.02-μm membrane before use. Two PLgel 5-μm mixed-c micrometer MIXED-C 300 × 7.5 mm columns (Polymer Laboratories Ltd., UK) were used with an eluent flow rate of 1.0 mL/min. The eluent from the columns passed through a UV detector (Knauer, Berlin, Germany) and then a LALLS cell followed by an RI cell. Both the LALLS and the RI detectors are combined in one integrated PL-LALS system (Polymer Laboratories). In this instrument the light source for both the LALLS and the RI detector is at the same wavelength (633 nm), which precludes any correction factor calculations as is usual with a white light source RI detector (4). The outputs from all three detectors were interfaced with a PL Caliber SEC workstation (Polymer Laboratories) for data collection and manipulation.

Narrow polydispersity diblock copolymers of PS–PMMA and PS–PEO were produced by anionic polymerization using conventional high-vacuum methods. The average AB copolymer composition was determined by H^1 NMR (model EM30, Varian, UK). Narrow dispersity PS and PMMA standards (Polymer Laboratories) were used for both instrument and SEC column calibrations. Samples were prepared as nominally 1-mg/mL solutions in the eluent and spiked with toluene as a flow rate marker before full loop 100-μL injection. Each copolymer was analyzed three times.

Results

PS–PMMA.   Figure 1 shows the typical raw data results obtained from the three detectors for the copolymer. A typical plot of the com-

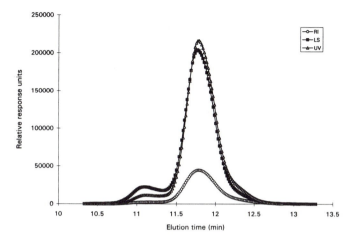

Figure 1.   Raw data chromatograms for PS–PMMA.

positional variation across the RI elution profile of the PS–PMMA copolymer calculated using equation 5 is shown in Figure 2. The excess
PS indicated at an elution time of around 12.4 min is considered to be
due to residual A block PS that was terminated when sampled from the
reactor or on the addition of the MMA. Based on a PS column calibration,
the molecular weight corresponding to this time is in excellent agreement
with the measured PS A block of 120,000 g/mol. It is of interest to note
that the early elution composition reflects the composition of the main
peak and indicates the presence of doubling up of species probably

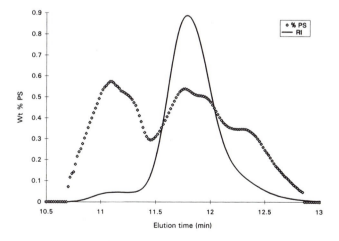

*Figure 2.   Compositional variation as a function of elution time (RI trace)
for PS–PMMA.*

occurring at termination. The average copolymer composition, calculated by using total peak areas in equation 5, was found to be 45.5% PS, 54.5% PMMA. This composition is in good agreement with the average composition calculated by NMR of 42.5% PS, 57.5% PMMA.

The SEC calibration data for both the PS and PMMA standards were fitted using a first-order polynomial

$$\text{Log } M = A + Bt$$

where M is peak molecular weight of the standard and t is the corresponding elution time. The coefficients for each calibration were as follows:

$$A_{PS} = 11.0860 \qquad B_{PS} = -0.4870$$
$$A_{PMMA} = 11.1466 \quad B_{PMMA} = -0.4873$$

The PS and PMMA equivalent M_w values calculated for the PS–PMMA copolymer (RI response) and the M_w found after applying the Runyon method to the UV and RI response values are shown in Table I. The PS equivalent M_w is lower than the PMMA equivalent M_w, and the copolymer M_w, as expected, is calculated to lie between these two.

The data analyzed from the UV–LALLS–RI chromatograms by applying equation 7 resulted in a calculated average ν for the copolymer of 0.1265 mL/g. This compares favorably with a theoretical value of 0.1275 mL/g calculated from equation 8, assuming ν for PS and PMMA in THF at 25 °C to be 0.185 and 0.085 mL/g, respectively (5) and a composition of PS to be 42.5%. This value was further verified by off-line differential refractometer measurements using a PL-DRI (Polymer Laboratories) from which a value of 0.1288 mL/g was obtained. Off-line multiangle classical light-scattering measurements were also performed on the same sample using a PL-LSP (Polymer Laboratories) that resulted in an M_w of 290,000 g/mol.

Table I. M_w Results (grams per mole) for PS–PMMA Copolymer

Method	1	2	3	Mean	% Var
PS equiv (RI, resp.)	258,638	258,964	260,042	259,210	0.23
PMMA equiv (RI, resp.)	288,428	288,791	290,054	289,090	0.24
UV–RI (Runyon)	274,244	274,481	275,881	274,870	0.26
UV–LALLS–RI (V)	283,749	282,952	281,999	282,900	0.25
UV–LALLS–RI (Vi)	273,291	271,976	273,276	272,850	0.23

M_w was calculated using the UV–LALLS–RI responses by applying equation 6 and assuming the average value of v calculated (0.1265 mL/g) across the whole distribution. M_w was then recalculated using vi values obtained via the $W_A i$ and $W_B i$ values at each slice. These two results are shown in Table I. The copolymer M_w obtained using the average value of v (282,900 g/mol) was somewhat higher than the copolymer M_w calculated using the Runyon dual detector approach (274,870 g/mol). However, when the vi values were applied, the calculated M_w of 272,850 g/mol was in good agreement with the dual detector approach. Thus, the triple detector approach correcting for vi at each elution volume gives better accuracy for the copolymer molecular weight where the composition varies across the MWD.

PS–PEO.   THF is not a preferred SEC eluent for PEO due to both solubility and adsorption problems (6). Therefore, the dual detector approach could not be applied in the case of the PS–PEO copolymer because an SEC calibration using PEO standards was not feasible. Figure 3 shows typical raw data chromatograms of the PS–PEO copolymer from the three detectors. In the triple detector approach, the detector constant $K_{PEO}ri$ was extrapolated from the $K_{PS}ri$ and $K_{PMMA}ri$ values assuming a v_{PEO} of 0.050 mL/g (5). Figure 4 shows the compositional variation of this sample across the elution profile. The average composition for this copolymer as determined by NMR was 96.2% PS, 3.8% PEO. With such a small amount of PEO, the copolymer M_w is likely to be quite similar to that of the starting A block PS, and the margin of error in the vi correction is likely to be higher. The PS equivalent M_w for the copolymer calculated from the RI response is shown in Table II. This value is con-

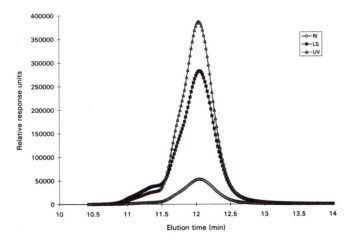

Figure 3.   Raw data chromatograms for PS–PEO.

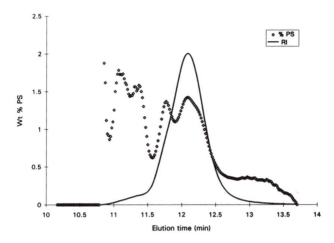

Figure 4.   Compositional variation as a function of elution time (RI trace) for PS–PEO.

siderably lower than that measured for the A block PS that was sampled during polymerization and found to have molecular weight of around 230,000 g/mol. This phenomena could be explained by two effects causing late elution after incorporation of the PEO resulting in lower molecular weight values calculated via conventional SEC calibration: adsorption of the copolymer and reduction in copolymer size (hydrodynamic volume). Theoretically, these phenomena should not affect the molecular weight values determined by SEC–LALLS.

The data analyzed from the UV–LALLS–RI chromatograms by applying equation 7 resulted in a calculated average ν for the copolymer of 0.1732 mL/g. This compares well with a theoretical value of 0.1799 mL/g assuming the values of ν_{PS} and ν_{PEO} to be 0.185 and 0.050 mL/g, respectively, (5) and a composition of 96.2% PS. The marginal difference could be attributed to error in the value assumed for ν_{PEO} or error in the copolymer composition determined by NMR. For example, a composition of 91.5% PS would give a theoretical ν of 0.1735 mL/g. However, in the absence of more experimental results, the original NMR composition has to be accepted.

Table II. M_w Results (grams per mole) for PS–PEO Copolymer

Method	1	2	3	Mean	% Var
PS equiv (RI, resp.)	192,363	196,560	198,067	195,660	1.23
UV–LALLS–RI (V)	261,942	262,382	263,787	262,700	0.30
UV–LALLS–RI (Vi)	228,920	231,291	227,118	229,110	0.75

The copolymer M_w was calculated using equation 6 by applying both the average ν (0.1723 mL/g) and the νi values obtained by the triple detection system. Both results, shown in Table II, are higher than the PS equivalent M_w calculated by conventional SEC column calibration and comparable with the A block PS molecular weight. This would suggest that the SEC–LALLS results are more reliable in this particular case for which secondary effects are evident. M_w determined using the average ν (262,700 g/mol) was in reasonable agreement with that determined by off-line multiangle classical light scattering (280,000 g/mol) in which methyl ethyl ketone was used as the solvent. The correction for νi resulted in a considerably lower value of M_w that may have some degree of error associated with it due to the small amount of comonomer present.

Summary

Multiple detection applied to the SEC characterization of copolymers is attractive because it yields both CCD and MWD information. A dual detection system based on two concentration detectors, for example, RI and UV, is useful where narrow standards of the homopolymers are available and where both homopolymers obey universal calibration. However, in other copolymer systems the addition of a third detector, LALLS, can offer the advantage of on-line determination of molecular weight for each eluting species. The triple detection approach gave similar values to the dual detector approach for a model copolymer system (PS–PMMA) studied. It was also applicable to a more difficult copolymer system (PS–PEO), although it appeared that where one homopolymer was present in very small quantities, an average ν value gave more consistent results than correction for νi across the distribution.

References

1. Meehan, E.; McConville, J. A.; Williams, A. G.; Warner, F. P. Presented at the 17th Annual Meeting of the Federation of Analytical Chemistry and Spectroscopy Societies, Cleveland, OH, October 1990.
2. Meehan, E.; McConville, J. A.; Williams, A. G.; Warner, F. P. Presented at the International GPC Symposium, Newton, MA, October 1989.
3. Runyon, J. R.; Barnes, D. E.; Rudd, J. F.; Tung, L. H. *J. Appl. Polym. Sci.* 1969, *13*, 2359.
4. Grinshpun, V.; Rudin, A. *J. Appl. Polym. Sci.* 1986, *32*, 4303–4311.
5. Huang, S. Presented at the 17th Annual Meeting of the Federation of Analytical Chemistry and Spectroscopy Societies, Cleveland, OH, October 1990.
6. Barth, H. G. In *Detection and Data Analysis in Size Exclusion Chromatography*; Provder, T., Ed.; ACS Symposium Series 352; American Chemical Society: Washington, DC, 1987; pp 29–46.

RECEIVED for review January 6, 1994. ACCEPTED revised manuscript November 28, 1994.

Use of a Gel Permeation Chromatography–Fourier Transform Infrared Spectrometry Interface for Polymer Analysis

James N. Willis[1] and L. Wheeler[2]

[1] Lab Connections, Inc., 5 Mount Royal Avenue, Marlborough, MA 01752
[2] Polymer Group, Exxon Chemical Company, Linden, NJ 07036

An interface between gel permeation chromatography (GPC) and Fourier transform infrared (FTIR) spectrometry has been developed. With this system it is possible to collect solvent free polymer deposition and to measure their infrared spectra as a function of molecular weight. The mobile phase from the GPC effluent is converted into an aerosol and removed using a pneumatic nozzle. The sample is collected on a Ge disc that rotates below the nozzle. After the sample is collected, the disc is transferred to an FTIR spectrometer where the infrared spectrum of the sample is collected. Normal GPC sample concentrations (0.1–0.25 wt/vol%) give sufficient sample for useable FTIR signals. All normal GPC solvents can be effectively removed, and the interface works with both low temperature and high temperature GPC applications.

T̲HE DETERMINATION OF COMPOSITIONAL CHANGES across the molecular weight distribution of a polymer is of considerable interest to polymer chemists. This information allows the chemist to predict the physical properties and ultimately the performance of the polymer. Several analytical techniques are of use in determining these properties. Mass spectroscopy, NMR, viscosity measurements, light scattering, and infrared (IR) spectroscopy all can be used to provide data in one form or the other about the compositional details sought. Each method has its place in the determination of the details of the structure of a polymer. IR spectroscopy, generically known as Fourier transform IR (FTIR)

0065–2393/95/0247–0253$12.00/0

spectroscopy, is used extensively for bulk polymer identification and gross structural analysis; however, to use FTIR to study compositional changes, it is necessary to separate the polymer into its various molecular weight components. Gel permeation chromatography (GPC) is the method of choice for carrying out this separation. Combining these two technologies offers the industrial polymer chemist a new approach to studying the details of samples. There have been several reports of high-performance liquid chromatography–GPC–FTIR interfaces (1–7), most of which were used for reverse-phase chromatographic applications. We discuss a new instrument based on the system described by Gagel and Biemann (5) in which the design is extended to cover GPC applications.

The use of FTIR in polymer analysis has been restricted to either flow-through IR cells or to preparative chromatography. Dekmezian et al. (7) presented work in which GPC–FTIR was used. Flow-through cells and preparative chromatography have limitations that reduce their utility to the chemist. Flow-through cells require an IR window in the solvent and useful sample IR absorption bands. Furthermore, the spectra must be taken as the sample moves through the IR beam, which limits the sensitivity of the method. GPC solvents, which include tetrahydrofuran (THF), toluene, and trichlorobenzene (TCB) generally prohibit obtaining the full IR spectra of the polymers. Preparative chromatography, on the other hand, while it eliminate some of the problems of the flow-through method, has limitations of its own. It requires separation by GPC, collection of the various fractions of interest, removing the solvents by evaporation, and then preparation of individual samples suitable for IR analysis. The process is time consuming, prone to sampling errors, and generally limited to high concentrations of samples. The nozzle described by Gagel and Biemann (5) was a simple nozzle assembly that removed the solvent from samples undergoing reverse-phase chromatographic separations and deposited the dry sample on a disc that then could be examined by IR spectroscopy. The approach has been commercialized and extended to GPC methods. In this study the utility of the method is discussed and results from a series of polymer samples are presented.

Experimental Details

The chromatograph used was either a Waters 150C (Waters, Milford, MA) high-temperature system with TCB as the solvent or a Waters 510 pump and a Rheodyne injector (Rheodyne, Cotati, CA) with THF as the solvent. The FTIR unit was a Nicolet 510P equipped with a DTGS detector (Nicolet Instrument Corp., Madison, WI). The FTIR bench was purged and equipped with software capable of continuously collecting spectra. Both TCB and THF were effectively removed from the sample. For TCB experiments, a

Table I.   High-Temperature Application

Chromatography	Spectrometer	Liquid Chromatography-Transform Polymer System

Waters 150C
Columns, 5
  microstyragel HT
Mobile phase-TCB
Injection volume,
  400 μL
Concentration, 0.3%
Flow rate, 1 mL/min.
Temperature, 145 °C
	Nicolet 510P
	Detector, DTGS
	Resolution, 8 cm^{-1}
	Scans, 64/set
		Transfer line temperature,
		  145 °C
		Nozzle flow, 100 μL/min
		Sheath gas temperature,
		  165 °C
		Sheath gas, nitrogen
		Disc speed, 10 deg/min
		Gas flow rate, 25 (nebulizer)
		  and 40 (sheath)

high-temperature cabinet and heated transfer line was used to transfer the GPC flow from the chromatograph to the collection nozzle. The experimental conditions are presented in Tables I and II.

No special hardware or software is required for either the chromatograph or the FTIR. Interface to the chromatograph is accomplished by diverting the sample–solvent flow from the end of the GPC column to the nozzle assembly via a flow divider. The flow divider allows the selection of a portion of the flow from the chromatograph to the nozzle. The quantity of flow depends on the viscosity of the mobile phase and the nature of the polymer being analyzed. The nozzle can evaporate up to 150 μL/min THF and TCB. The nozzle design is shown in Figure 1.

The sample–solvent flow is mixed with a controlled flow of nitrogen or air that generates an aerosol. The aerosol flows through a needle and exits the nozzle. As it exits, it is surrounded by a heated sheath gas, again nitrogen or air, that serves to confine the spray to a 2.5–3-mm spot and provides the necessary energy to evaporate the mobile phase. The resultant dry sample is directed onto a 60-mm-diameter Ge disc that is placed on a rotating platform 5–10 mm below the nozzle. Disc rotation in the system is automatically started after injection of the sample. The main portion of each collection took ~10 min.

Table II.   Low-Temperature Application

Chromatography	Spectrometer	Liquid Chromatography-Transform Low Temp System

Waters 510 pump
Columns, 1 styragel
Mobile phase, THF
Injection volume, 100 μL
Concentration, 0.25%
Temperature, RT
	Nicolet 510P
	Detector, DTGS
	Resolution, 4 cm^{-1}
	Scans, 16–32/set
		Sheath gas temperature,
		  55 °C
		Flow rate, 60 μL/min
		Disc speed, 10 deg/min
		Sheath gas, nitrogen
		Gas flow rate, 25 (nebulizer)
		  and 40 (sheath)

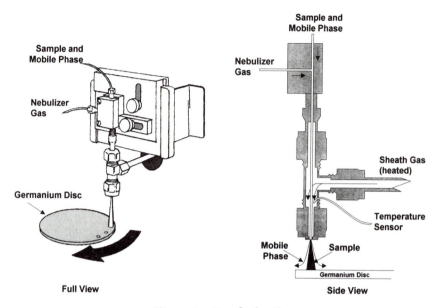

Figure 1. Nozzle detail.

Disc rotation can be automatically triggered from the GPC upon injection or at some preset delay so sample collection is virtually unattended. The speed of rotation of the disc is adjusted to match the time of the evolution of the sample from the chromatograph. At a speed of 10 deg/min, 36 min of sample collection can take place on each disc. The Ge discs are easily cleaned and can be reused. After collection, the disc is removed and placed in a 3X beam condenser within the FTIR optics cabinet. The beam condenser is designed to provide an optical match between the FTIR beam size, which is ~9 mm, and the size of the deposited sample, ~3 mm. Once seated on the platform in the beam condenser, the disc is rotated beneath the IR beam and spectra of individual samples collected. If the FTIR system has the capability of continually collecting spectra, then the spectra of the polymer deposit can be displayed continuously, thus generating an IR chromatogram. If this software is not available, the spectra may be individually collected and displayed by scanning the disc from point to point.

The concentration required for satisfactory analysis was between 0.1 and 0.5% by weight, assuming an injection volume of between 400 and 150 μL, respectively. These conditions are consistent with normal methods used in GPC analysis. At this concentration and assuming a distribution of the sample over a 20-min time period, a signal level of between 0.1 and 0.5 AU is observed on the FTIR spectrum. An example of a typical set of spectra is shown in Figure 2. The signal-to-noise ratio in all spectra are quite adequate for qualitative interpretation.

Starting conditions for each sample are easily obtained for most polymers; however, in some samples the nozzle exhibited instability that resulted in "splitting" rather than a smooth flow of polymer deposit onto the disc. This was correctable by adjusting the mix of sheath gas–neblulizer gas or by changing the temperature of the sheath gas.

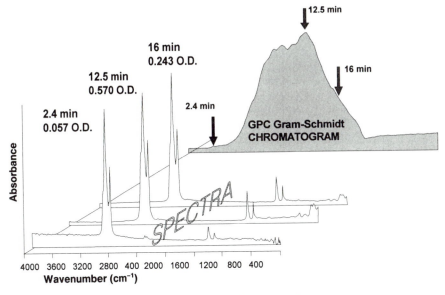

Figure 2.   Analysis of the ethylene–propylene copolymer.

Unless the THF used was obtained from either a fresh bottle or was stabilized with butylated hydroxytoluene (BHT), a constant stream of polymerized THF was observed as a light, white stripe of material. The presence of this material was a minor irritant and could be ratioed out of the spectrum in postcollection data processing.

Results and Discussion

High-Temperature Application. *Vinyl Acetate Distribution in Copoly(ethylene-vinyl acetate).*   In the characterization of polymers, molecular distribution and composition are two critical parameters. Every physical property and processing change of the material can be related to these two parameters. With copolymers, IR spectroscopy can be used for determination of the distribution of one or both monomers within the molecular weight distribution.

Determining vinyl acetate distribution in two poly(ethylene vinyl acetate) copolymers is a good illustration using the liquid chromatography–transform interface. Two samples were used, one with reported high vinyl acetate content and the other low. Both were made in an unbaffled autoclave with a single vinyl acetate injection point and are typical of copolymers used in heavy wall plastic bags.

Figure 3 shows the IR chromatogram (total CGM) of EVA-high and EVA-low spectra. Both peaks seem to have uneven shapes, which is in part due to instability in the nozzle. Individual spectra from EVA-high are shown in Figure 4 and those for EVA-low in Figure 5. Only the

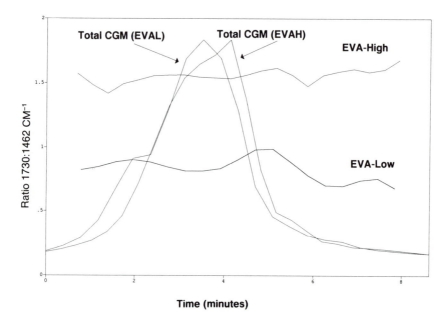

Figure 3.   EVA–PE distribution.

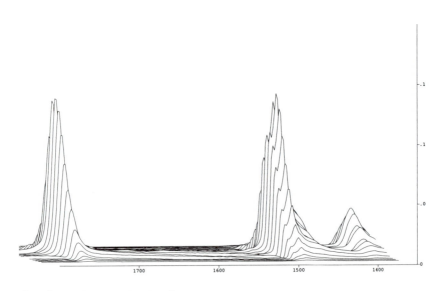

Absorbance–Wavenumber (cm⁻¹)

Figure 4.   3D IR spectra of EVA (high)–PE.

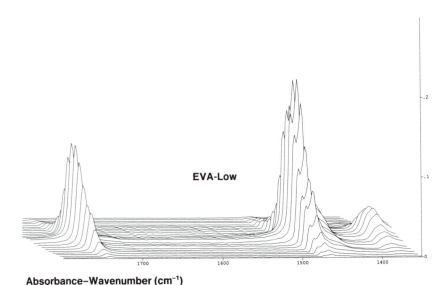

Absorbance–Wavenumber (cm⁻¹)

Figure 5.   3D IR spectra of EVA (low)–PE.

region from 1800 to 1350 cm⁻¹ is shown to better present the region containing both the band arising from the carbonyl group at 1730 cm⁻¹ from the vinyl portion of the polymer and from the band at 1462 cm⁻¹ from the C–H₂ backbone of the polymer. In EVA-high one can see the general appearance of the bands being approximately equal in strength from high molecular weight range to low. In EVA-low the C=O band is significantly lower in intensity. This change in the ratio of C=O to C–H₂ is a clear indicator of the changes in concentration of vinyl acetate. To obtain a quantitative measure of the concentration of the vinyl acetate within the copolymer, one would need to prepare a calibration curve in which a known amount of vinyl acetate was added.

   To determine the distribution of vinyl acetate within the samples, the ratio of the 1730–1462 cm⁻¹ band was measured across the molecular weight distribution. The resultant plots for EVA-high and -low are also shown in Figure 3. The ratio of EVA-low suggests that the concentration of vinyl acetate is uniform across the middle of the deposit and shows a slight decrease toward lower molecular weight. On the other hand, when the data from EVA-high was plotted, an even distribution of vinyl acetate was found. Some contribution from the C–H₂ from the vinyl acetate is expected but should be small relative to the large contribution from the polyethylene portion of the copolymer; therefore, no attempt was made to account for its contribution. The data suggest that the manufacturer has produced a copolymer that has a different distribution of vinyl acetate in one copolymer than the other. This difference affects the physical properties of the two products.

High-Density Polyethylene. One of the more difficult tasks with which analytical polymer chemists have been faced has been the characterization of the very simple molecules, polyethylene and polypropylene. From an IR spectroscopy point of view, there are very few bands to work with; the interesting information lies in subtle shifts in the weak bands and in intensity changes in one band compared with another. Once understood, however, these bands can be used to monitor such important characteristics as branching distribution and crystallinity in the polymer. Using a sample of polyethylene, NBS1475, which is a high-density polyethylene, a study was made of the changes that could be observed as a function of molecular weight.

Shown in Figure 6 are the spectra of the polymer in the C–H stretching region collected as a function of molecular weight. One of the features indicative of the branching is the intensity of the bands arising from the end group methyl vibrations. The higher the intensity of these bands, the more end groups are present, which in turn is an indication of the branching of the polymer. In Figure 7, one can clearly see a broadening of the entire band structure on the deposit as the molecular weight decreases from high (20–25 min) to low (40–45 min). The change can be used to monitor contribution of the end groups or branching in the sample.

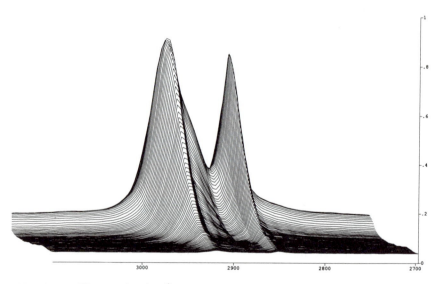

Absorbance–Wavenumber (cm⁻¹)

Figure 6.   3D IR spectra of high-density polyethylene (NBS1475).

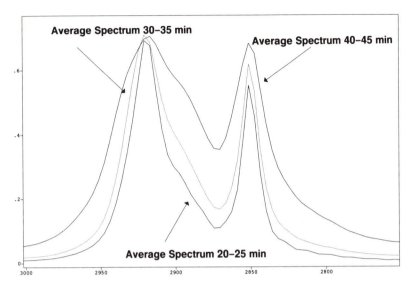

Absorbance–Wavenumber (cm⁻¹)

Figure 7.   Average spectra of high-density polyethylene (NBS1475).

Low Temperature Applications.   The versatility of the system in the determination of complex compositional changes is illustrated in a study of a jet oil lubricant. This sample is one of many that have been studied in which complex changes are found as one examines the various molecular weight fractions collected. The ability to easily determine the presence or absence of a particular component within a complex mixture is sometimes vital to the polymer chemist.

Figure 8 is a three-dimensional plot of the IR spectra collected from the polymer deposit on the disc. This presentation gives one an appreciation of the complexity of an industrial polymer sample, but points out that there is sufficient detail throughout the sample to distinguish features that can be used to identify individual components within the sample. At the beginning of the deposit, the ester carbonyl frequency occurs at 1733 cm^{-1} and gradually shifts to a higher frequency toward the low molecular weight end of the deposit, ending at 1742 cm^{-1}. This indicates a change in composition or in the species present. In the early part of the deposit, there is a band at 3200 cm^{-1} that is indicative of an organic acid. Bands at 3000–3100 cm^{-1} arise from the C–H stretching modes of an aromatic ring. The strong band as 1230 cm^{-1} suggests the presence of an ether linkage. These data taken together indicate that at least part of the initial deposit contains a phenolic based ester and an acid. The phenyl vibrations gradually disappear toward the end of the deposit, leaving a completely new spectra of what appears to be still an

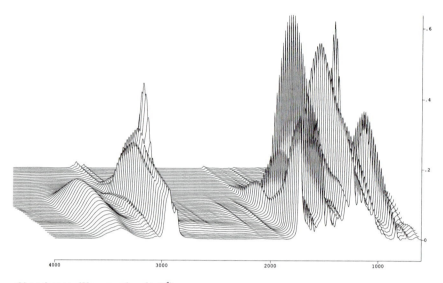

Absorbance–Wavenumber (cm⁻¹)

Figure 8.   3D IR spectra of jet oil lubricant.

ether, but this has more aliphatic character. This information is consistent with the known structures of lubricants and oils, and with the use of library search routines one could confirm the suggested structures. The fact that the GPC did not completely separate the polymer into its individual components does not prevent the chemist from isolating the components within this complex sample.

Conclusions

The combination of GPC and FTIR offers the polymer chemist the possibility of examining a variety of materials for compositional distribution, presence or absence of functional groups, and branching content without extensive sample preparation and without integrating the chromatography and spectroscopy laboratory. The interface performs well with both room temperature and high-temperature GPC applications and provides adequate amounts of sample for qualitative IR studies.

References

1. Jinno, K.; Fujimoto, C.; Ishii, D. *J. Chromatogr.* **1982**, *239*, 625.
2. Kuehl, D.; Griffiths, P. R. *J. Chromatogr. Sci.* **1979**, *17*, 471.
3. Griffiths, P. R.; Conroy, C. M. *Adv. Chromatogr.* **1986**, *25*, 105.

4. Kalasinsky, V. F.; Whitehead, K. G.; Kenton, R. C.; Smith, J A. S.; Kalasinsky, K. *J. Chromatogr. Sci.* **1987,** *25,* 273.
5. Gagel, J. J.; Biemann, K. *Anal. Chem.* **1987,** *59,* 1266.
6. Lange, A. J.; Griffiths, P. R.; Fraser, D. J. J. *Anal. Chem.* **1991,** *63,* 782.
7. Dekmezian, A. H.; Morioka, T.; Camp, C. E. *J. Polym. Sci., Polym. Phys. Ed.* **1990,** *28,* 1903.

RECEIVED for review January 6, 1994. ACCEPTED revised manuscript July 8, 1994.

Size-Exclusion Chromatography–
Fourier Transform IR Spectrometry
Using a Solvent-Evaporative Interface

P. C. Cheung,[1] S. T. Balke,[1] and T. C. Schunk[2]

[1] Department of Chemical Engineering and Applied Chemistry, University of Toronto, Toronto, Ontario M5S 1A4, Canada
[2] Analytical Technology Division, Research Laboratories, Eastman Kodak Company, Rochester, NY 14650–2136

A solvent-evaporative interface is used to deposit each fraction obtained from size-exclusion chromatography (SEC) as a dry polymer film on an IR transparent disc for subsequent IR analysis. Spectra without solvent interference bands result. However, in addition to removing the solvent, the interface must provide polymer films that yield undistorted spectra. Christiansen distortion is particularly troublesome because it interferes with quantitative interpretation in the affected spectrum. Undistorted spectra were invariably obtained from continuous films and sometimes obtained from discontinuous films (separate particles). Carbon-coated KCl discs and either bare or carbon-coated germanium discs provided morphologies with good spectra. Polymer deposition on bare KCl discs often provided unacceptable morphologies. Surface-wetting properties of the substrate appear to dominate deposit morphology.

A FOURIER TRANSFORM INFRARED (FTIR) SPECTROMETER is potentially a very powerful detector for size-exclusion chromatography (SEC). For polyolefins in particular, SEC–FTIR may provide for each molecular size present in the sample, measurements of such properties as degree of unsaturation, branching frequency, and chemical composition of copolymers. Although FTIR spectrometers have already been used as SEC detectors by employing micro-sized flow cells (1), this method is constraining: Very few spectral windows are available in commonly used mobile phases, and decreasing path length opens windows but simul-

0065–2393/95/0247–0265$12.00/0

taneously reduces polymer signal. An alternative to the use of a flow cell is removal of the solvent from SEC fractions followed by IR analysis of each of the dried films. Preparative SEC followed by collection of fractions and mobile-phase evaporation is impractically slow and costly for routine use. An on-line evaporative interface can accomplish solvent removal during the SEC run. In this work, we employ an interface design based upon the one developed by Dekmezian et al. (2), for high-temperature SEC. In recent assessments of this interface and a comparison of it with a commercially available room-temperature evaporation interface using the Gagel and Biemann design (3–5), it was found, for both devices, that the morphology of the deposited film critically affected the resulting IR spectra. The term "film morphology" here refers to the structure of the polymer film on the scale of mid-IR wavelengths. Film morphology, the focus of this chapter, is currently the primary obstacle to the use of either interface for quantitative IR detection.

Theory

Considerations for High-Temperature SEC Analysis of Polyolefins. In the use of a solvent-evaporative interface with high-temperature SEC, three primary complications are involved: the low volatility of the 1,2,4-trichlorobenzene (TCB) mobile phase, differing polymer solubilities in TCB, and the semicrystalline nature of polyolefins. The boiling point of TCB is 213 °C at 101.3 kPa (1 atm), and the latent heat of vaporization is 48 kJ/mol. Thus to reduce operating temperature, reduced pressure is utilized to assist evaporation. Solubilities (and hence tendency to phase-separate in a sprayed droplet) of polyolefins depend upon molecular weight and temperature. Whereas polypropylene is generally less soluble than polyethylene at elevated temperatures, other polymers (e.g., polystyrene) are soluble even at room temperature. Polymer blends and copolymers would be expected to display film morphologies reflecting different rates of phase separation during drying. Crystallinity will also affect film morphology and varies with polymer type as well as with degree of branching.

Film Properties Affecting IR Spectra. In addition to film thickness, influential properties include uniformity, wedging, dispersion characteristics, molecular interactions, and degree of crystal orientation (6). Film nonuniformity can cause offsets in the absorption-band intensities of the spectrum. A wedge or sloping thickness sample profile can lead to photometric error (7). The dispersion effect, also known as the Christiansen effect, is caused by the scattering of radiation. The extent of scattering depends on the particle size and refractive index differences (8–10). Christiansen scattering may not be significant when there are particulates with diameters significantly smaller than the IR wavelength. Scattering can cause both baseline curvature and derivative-like, absorption-band shape changes. Molecular interactions can cause shifts of absorption bands. Polarization effects may lead to changes in relative intensities in a spectrum due to crystal orientation.

Film Morphology Effects Observed in Previous Interface Investigations. As mentioned earlier in both previous studies (*3, 4*) film morphology effects sometimes strongly affected the IR spectra. The presence of strong derivative-like bands (Christiansen effect) as well as highly sloping baselines frequently occurred. The latter problem was considered much less serious than the former because the absorption bands involved were often quite narrow (baselines could be assumed linear under the peak). In contrast when the Christiansen effect occurred, no useful quantitative information could be obtained from the part of the absorption spectrum affected. Electron and optical microscopy revealed that morphologies displaying multiple isolated particles of 1–15 μm provided particularly distorted spectra. In one demonstration of the importance of film morphology, exposing a film of polystyrene to a solvent vapor completely removed the sloping baseline by forming a continuous film (*3, 4*).

Issues. Interpretation of the Christiansen distorted portion of spectra is not generally practical. Generation of film morphologies that do not adversely affect the IR spectra is the requirement. With the limited goal of accomplishing reliable quantitative analysis using a solvent-evaporative interface, it is necessary to further elucidate several interrelationships: the effect of film morphology on the IR spectra (sufficient to disclose those morphological properties causing spectral distortions), the effect of polymer type on film morphology (sufficient to reveal the material's contribution to the issue), and the effect of operating conditions on film morphology (sufficient to permit acceptable morphologies to be reliably obtained). Complexities include interactions among polymer type and operating conditions.

The work progressed in two phases: In Phase I, the objective was to determine the causes for film morphologies that resulted in the Christiansen distortion; in Phase II, the objective was to use the knowledge gained from Phase I to devise methods of experimentally eliminating these causes.

Experimental Details

The Dekmezian solvent-evaporative interface design consists of an ultrasonic nozzle installed in a heated vacuum chamber. The mobile phase is spray-dried over discs that are sequentially placed below the nozzle in a programmed manner. At the end of the run, a wheel containing the discs is removed for IR analyses. Our modifications to this design included a higher frequency ultrasonic nozzle (120 versus 60 khz), a cooling jacket on the nozzle, and placement of a heater between the nozzle and disc (Phase I) or below the wheel containing discs (Phase II). Figure 1 shows a schematic of the interface in Phase II. Further details may be found elsewhere (3). The interface was at the outlet of a model 150C high-temperature size-exclusion chromatograph (SEC) (Waters Associates, Milford, MA).

In both Phases I and II of the work, the SEC was equipped with three PLgel (Polymer Laboratories, Amherst, MA) 10-μm mixed-bed analytical columns. TCB was the mobile phase. Injection volume was 100 μL of each 0.2 wt% polymer sample dissolved in TCB with 0.2 wt% butylated hydroxytoluene as a stabilizer. Flow rates were generally 0.5 mL/min. A Mattson Galaxy 6020 FTIR spectrometer and a Mattson Quantum infrared microscope (Madison, WI) equipped with mercury cadmium telluride (MCT) detectors as well as a Nikon SMZ-2T optical microscope was used to analyze the films. The IR spectra shown in this study were obtained by averaging 128 scans at 4-cm^{-1} resolution under transmission mode and were not baseline corrected. Polypropylene (PP 180K, American Polymer Standards, Mentor, OH), polystyrene (NBS 706), and linear (high-density) polyethylene (NBS 1475, NIST, Washington, DC) were analyzed individually and combined pairwise in equal weights.

In Phase I, the columns were bypassed for some of the runs. The vacuum oven was at 153–160 °C and 10–30 kPa. KBr discs (13 mm in diameter and

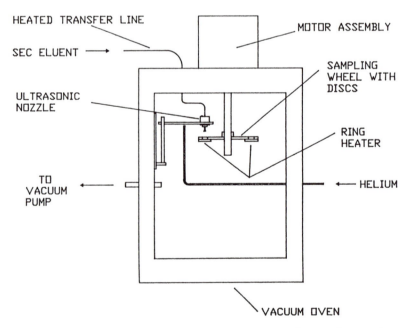

Figure 1. A schematic of the solvent-evaporative interface in Phase II. (Reproduced with permission from reference 3. Copyright 1993 John Wiley and Sons.)

2 mm thick) were used to collect polymer deposit. The temperature of the spray close to the KBr disc was 180–190 °C, and nozzle power ranged from 0.3 to 0.8 W.

In Phase II, the vacuum oven was at 102–108 °C and 5–7 kPa. KCl discs (13 mm in diameter and 2 mm thick) and germanium discs (13 mm in diameter and 1 mm thick) were used to collect polymer deposit. Some discs were vapor coated with carbon by an Edwards coating system E306A (United Kingdom). The temperature at the heater surface was 185–215 °C and nozzle power ranged from 0.2 to 0.4 W.

The criteria for selecting experimental conditions (except in one case mentioned later where liquid accumulation was desired) were to ensure no boiling of liquid at the tip of the nozzle and no liquid accumulation on the disc. The observed morphologies did not change significantly within the range of operating conditions listed for each phase of work.

Results and Discussion

Phase I: Causes for Undesired Film Morphologies. *The Effect of Film Morphology on IR Spectra.* As in earlier studies (3, 4), the two major spectral distortions observed were derivative spectra and sloping baselines. However, in this case, a large number of high-quality spectra were also observed. Figure 2 shows the worst spectrum obtained and the accompanying "dispersed particle" film morphology. It represents a SEC analysis of a polyethylene sample. Figure 3 shows another analysis

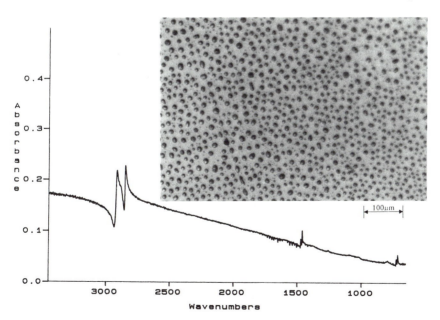

Figure 2. IR spectrum of polyethylene and film morphology (inset) for discontinuous film.

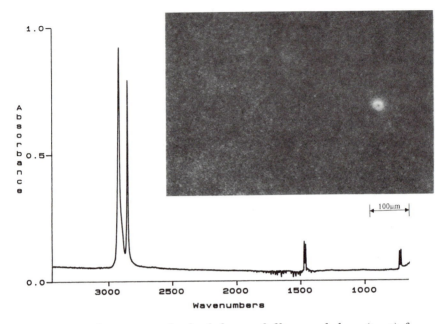

Figure 3. IR spectrum of polyethylene and film morphology (inset) for continuous film.

of polyethylene. In this case, the spectrum was excellent and the film morphology appeared to be a continuous film. Figure 4 shows an analysis of a polymer blend of polyethylene with polystyrene. The film morphology appears complex but, again, a continuous film is evident with no areas devoid of polymer and the spectrum is excellent. Examination of the many photomicrographs obtained in this study revealed progressive improvement in the spectra as the bare areas in the deposit were filled in by polymer. Apparently, the single most important requirement for an acceptable spectrum is a continuous film. The presence of irregularities in the film or even significant changes in polymer composition (and hence refractive index) were of much less importance. Furthermore, the appearance of the deposits leads to the hypothesis that the polymer must wet the surface of the KBr disc to provide consistent acceptable film morphologies.

When a continuous film was obtained, no Christiansen effect was observed. However, sometimes good spectra were obtained from discontinuous films. The photomicrographs and IR spectra shown in Figure 5 were taken using an IR microscope. For the polyethylene deposit consisting of droplets 10–30 μm in diameter, no spectral distortion was observed. In another deposit from the same SEC run when droplets ranged from 2 to 10 μm in diameter, a distorted spectrum was obtained.

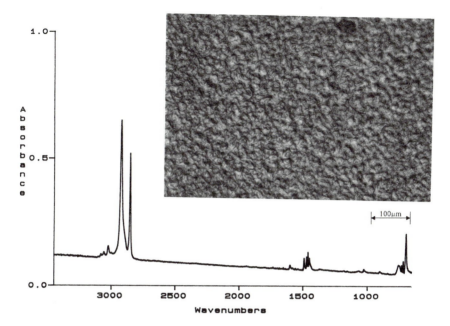

Figure 4.   IR spectrum of polyethylene–polystyrene blend and film morphology (inset) for continuous film.

Thus, when a discontinuous film was present, the size and possibly the shape of the deposited particles determined whether or not the Christiansen effect was observed.

The Effect of Polymer Type on Film Morphology.   At relatively high polymer concentrations, all types of polymers used, and their blends formed continuous films and provided excellent spectra. Blend composition and the presence or absence of crystallization did not cause Christiansen effect distortions. However, at concentrations usual for SEC analysis, all polymers showed evidence of not wetting the KBr surface, yielded discontinuous films, and often had poor spectra. These results indicate, with respect to spectral distortions, that the main importance of polymer type is its influence on continuous-film formation.

The Effect of Evaporation Conditions on Film Morphology.   Two extreme modes of operation were examined: evaporation sufficiently rapid to avoid any accumulation of solution on the surface of the KBr disc during the run, and evaporation so mild that liquid accumulated on the disc surface and then evaporated completely. Evaporation control variables included ultrasonic nozzle power, spray temperature, gas flow rate around the disc, oven temperature, and pressure. With accumulation

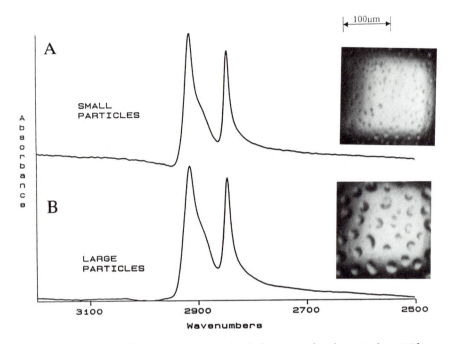

Figure 5.   Micro-IR spectra of the polyethylene samples deposited on KCl discs shown in photomicrograph insets. Key: A, particle sizes of 2 to 10 μm; and B, particle sizes of 10 to 30 μm.

of liquid on the disc, the deposited film morphology was heterogeneous, and, in some cases, polymer was transported away from the disc's center. For high evaporation rates (no liquid accumulation), the polymer particles were smaller and the morphology was more homogeneous. Morphology may be very sensitive to many different variables. From a practical viewpoint, the objective is to gain sufficient control to prevent unacceptable morphologies from forming. If the observation that a continuous film is a sufficient condition for undistorted spectra is correct, then this becomes the key objective.

Phase II: Obtaining Acceptable Film Morphologies.   Hypothesizing that the failure of the polymer solution to wet the surface of the substrate was the root cause of discontinuous films, we therefore made attempts to modify the wetting properties of the substrates used in the solvent-evaporative interface. This modification was done by using vapor deposition to coat a thin film of carbon (less than 0.01 μm) on the substrates. Both KCl discs and germanium discs were carbon-coated. Contact angle measurements revealed that with the carbon coating the TCB solution completely wetted both substrates. Without the coating, TCB did not wet either substrate. To evaluate the effect of the surface modification in SEC–FTIR analysis, some substrates were partially masked during the carbon-deposition process to provide a disc where only half of the surface was carbon-coated.

On KCl discs, the carbon coating dramatically changed both the morphology and spectra obtained. Figure 6 shows polyethylene deposited on a disc surface that was half-coated by carbon. The noncoated half shows isolated polymer particles, whereas the carbon-coated half shows coalesced (although still isolated) particles. The corresponding spectra show that, unlike the morphology on the noncoated surface, carbon coating provided a morphology that did not produce distortion and increased band intensity. Figures 7 and 8 show similar results for the deposition of polypropylene and the deposition of a polyethylene–polypropylene blend. In every case the deposition on the carbon-coated surface provided acceptable spectra. The blend deposit appeared as a continuous film, whereas those of the individual polymer components of the blend did not. Furthermore, the carbon-coated surface of the disc contained a noticeably larger amount of polymer resulting in greater absorbance band intensity.

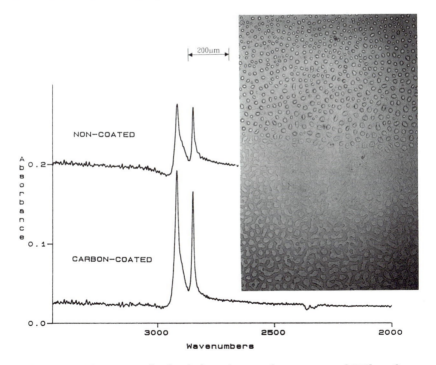

Figure 6.   IR spectra of polyethylene deposited on noncoated KCl surface and carbon-coated KCl surface with corresponding photomicrograph. The top half of the photo corresponds to the noncoated surface and the bottom half to the carbon-coated surface.

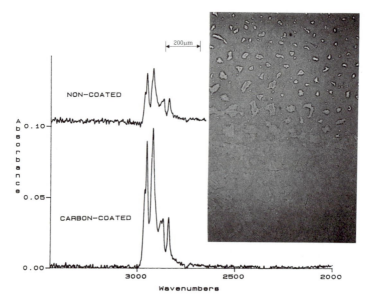

Figure 7. IR spectra of polypropylene deposited on noncoated KCl surface and carbon-coated KCl surface with corresponding photomicrograph. The top half of the photo corresponds to the noncoated surface and the bottom half to the carbon-coated surface.

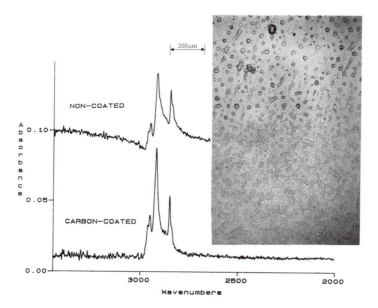

Figure 8. IR spectra of polyethylene–polypropylene blend deposited on noncoated KCl surface and carbon-coated KCl surface with corresponding photomicrograph. The top half of the photo corresponds to the noncoated surface and the bottom half to the carbon-coated surface.

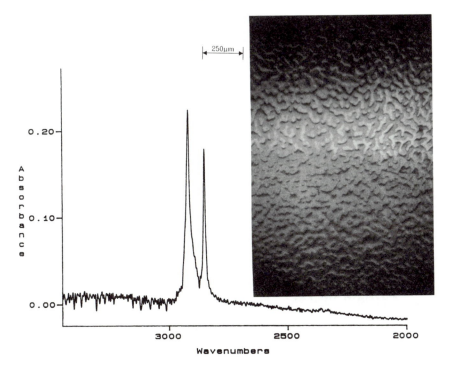

Figure 9.   IR spectrum of polyethylene deposited on noncoated germanium surface with the corresponding photomicrograph on a germanium disc half-coated with carbon. The top half of the photo corresponds to the noncoated surface and the bottom half to the carbon-coated surface.

Germanium was substituted for KCl with the idea of obtaining more rapid evaporation of any solvent reaching the surface of the disc. Germanium has a thermal conductivity that is an order of magnitude greater than that of KCl. With a disc heater at the base of the disc during deposition, it would be expected that evaporation would be much more rapid from the germanium disc than from the KCl disc. Figures 9 through 11 show the results with germanium. In every case the carbon coating had no effect on the deposited morphology and good spectra were obtained from both the carbon-coated part of the disc and from the noncoated part. Only the IR spectra on the noncoated part are shown in the figures, and those spectra obtained on the carbon-coated part were similar. Also, morphologies obtained appeared very similar to those obtained on the corresponding carbon-coated part of the KCl discs (compare morphologies in Figures 6 and 9, 7 and 10, and 8 and 11).

To investigate the contribution of evaporation rate on the deposit morphologies on germanium discs, an experiment was conducted at a low evaporation rate using a polyethylene–polypropylene blend on one

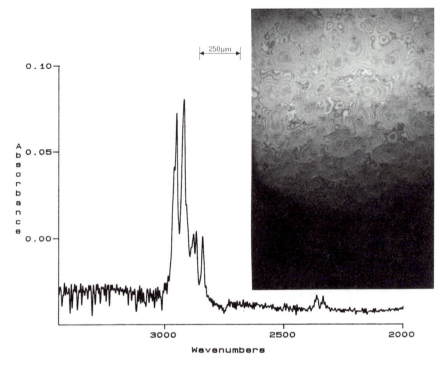

Figure 10.   IR spectrum of polypropylene deposited on noncoated germanium surface with the corresponding photomicrograph on a germanium disc half-coated with carbon. The top half of the photo corresponds to the noncoated surface and the bottom half to the carbon-coated surface.

half carbon-coated disc. The resulting morphology is shown in Figure 12. The deposit did show familiar large-scale "waves" of deposited polymer (not apparent under magnification) because of liquid accumulation on the disc prior to evaporation (3). However, the important observation was that no differences between the morphology on the carbon-coated and the noncoated part of the disc were evident.

One possible explanation for these results and the observation that TCB did not wet the bare germanium disc is that it is the wetting properties of the molten polymer that are important and not that of TCB. The temperature of the surface of all discs used exceeded the melting points of all the polymers investigated for the entire experiment. Furthermore, with the evaporation conditions used for all experiments except the last one mentioned, evaporation of solvent was instantaneous when the droplets reached the disc surface. No liquid accumulation was evident.

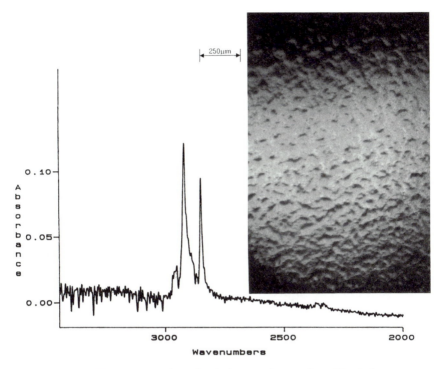

Figure 11. IR spectrum of a polyethylene–polypropylene blend deposited on noncoated germanium surface with the corresponding photomicrograph on a germanium disc half-coated with carbon. The top half of the photo corresponds to the noncoated surface and the bottom half to the carbon-coated surface.

Conclusions

Phase I work revealed that a sufficient condition for obtaining spectra without the Christiansen effect distortion was that the deposited film has no areas bare of polymer. When such a continuous film was obtained, other microstructural features and even polymer type were of secondary importance and were possibly even unimportant in obtaining good quality spectra. It was possible to obtain undistorted spectra from discontinuous films. However, particle size range of the deposit appeared critical to the results. The diversities of polymer morphologies possible and their sensitivity to operating conditions meant that tailoring particle size would be very difficult. Deposition control to ensure continuous films appeared as a much more attainable objective and was pursued in Phase II.

In Phase II, two methods of forming continuous films were examined: modification of the wetting properties of the substrate surface and flash evaporation from the substrate. Wetting properties of the surface were

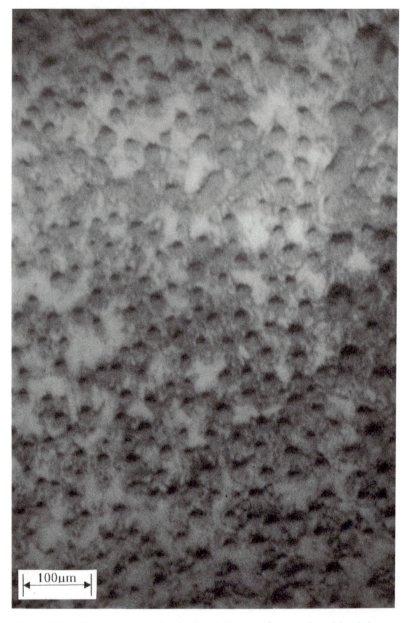

Figure 12. Photomicrograph of polyethylene–polypropylene blend deposited at a low evaporation rate on a germanium disc half-coated with carbon. The top half of the photo corresponds to the noncoated surface and the bottom half to the carbon-coated surface.

shown to be the important factor. Deposits on carbon-coated KCl and deposits on either bare or carbon-coated germanium provided morphologies yielding acceptable spectra.

Acknowledgments

We are pleased to acknowledge that this study was funded by grants from Eastman Kodak Company, Rochester, NY, the Ontario Centre for Materials Research, and the Natural Sciences and Engineering Research Council of Canada.

References

1. Hellgeth, J. W.; Taylor, L. T. *Anal. Chem.* **1987**, *59*, 295–300.
2. Dekmezian, A. H.; Morioka, T.; Camp, C. E. *J. Polym. Sci., Polym. Phys. Ed.* **1990**, *28*, 1903–1915.
3. Cheung, P.; Balke, S. T.; Schunk, T. C.; Mourey, T. H. *J. Appl. Polym. Sci.: Appl. Polym. Symp.* **1993**, *52*, 105–124.
4. Schunk, T. C.; Balke, S. T.; Cheung, P. *J. Chromatogr. A,* **1994**, *661*, 227–238.
5. Gagel, J. J.; Biemann, K. *Mikrochim. Acta* **1988**, *11*, 185–187.
6. Hannah, R. W. In *Advances in Applied Fourier Transform Infrared Spectroscopy;* Mackenzie, M. W., Ed.; John Wiley and Sons: New York, 1988; pp 1–42.
7. Hirschfeld, T. *Anal. Chem.* **1979**, *51*, 495–499.
8. Smith, A. L. *Applied Infrared Spectroscopy: Fundamentals, Techniques, and Analytical Problem-Solving;* John Wiley and Sons: New York, 1979; p 78.
9. Prost, R. *Clays Clay Miner.* **1973**, *21*, 363–368.
10. Kendall, D. N. In *Applied Infrared Spectroscopy;* Kendall, D. N., Ed.; Reinhold: New York, 1966; pp 137–138.

RECEIVED for review January 6, 1994. ACCEPTED revised manuscript August 10, 1994.

INDEXES

Author Index

Affiliation Index

Subject Index

Production: Margaret J. Brown
Acquisition: Anne Wilson
Indexing: Gloria R. Hamilton
Cover design: Neal R. Clodfelter

Typeset by Tapsco, Akron, PA
Printed and bound by Maple Press, York, PA

Highlights from ACS Books

Other ACS Books

Biotechnology and Materials Science: Chemistry for the Future
Edited by Mary L. Good
160 pp; clothbound, ISBN 0–8412–1472–7, paperback, ISBN 0–8412–1473–5

Chemical Demonstrations: A Sourcebook for Teachers
Volume 1, Second Edition by Lee R. Summerlin and James L. Ealy, Jr.
192 pp; spiral bound; ISBN 0–8412–1481–6
Volume 2, Second Edition by Lee R. Summerlin, Christie L. Borgford, and Julie B. Ealy
229 pp; spiral bound; ISBN 0–8412–1535–9

The Language of Biotechnology: A Dictionary of Terms
By John M. Walker and Michael Cox
ACS Professional Reference Book; 256 pp;
clothbound, ISBN 0–8412–1489–1; paperback, ISBN 0–8412–1490–5

Cancer: The Outlaw Cell, Second Edition
Edited by Richard E. LaFond
274 pp; clothbound, ISBN 0–8412–1419–0; paperback, ISBN 0–8412–1420–4

Chemical Structure Software for Personal Computers
Edited by Daniel E. Meyer, Wendy A. Warr, and Richard A. Love
ACS Professional Reference Book; 107 pp;
clothbound, ISBN 0–8412–1538–3; paperback, ISBN 0–8412–1539–1

Practical Statistics for the Physical Sciences
By Larry L. Havlicek
ACS Professional Reference Book; 198 pp; clothbound; ISBN 0–8412–1453–0

The Basics of Technical Communicating
By B. Edward Cain
ACS Professional Reference Book; 198 pp;
clothbound, ISBN 0–8412–1451–4; paperback, ISBN 0–8412–1452–2

The ACS Style Guide: A Manual for Authors and Editors
Edited by Janet S. Dodd
264 pp; clothbound, ISBN 0–8412–0917–0; paperback, ISBN 0–8412–0943–X

Personal Computers for Scientists: A Byte at a Time
By Glenn I. Ouchi
276 pp; clothbound, ISBN 0–8412–1000–4; paperback, ISBN 0–8412–1001–2

Chemistry and Crime: From Sherlock Holmes to Today's Courtroom
Edited by Samuel M. Gerber
135 pp; clothbound, ISBN 0–8412–0784–4; paperback, ISBN 0–8412–0785–2

For further information and a free catalog of ACS books, contact:
American Chemical Society
Distribution Office, Department 225
1155 16th Street, NW, Washington, DC 20036
Telephone 800–227–5558

Bestsellers from ACS Books

The ACS Style Guide: A Manual for Authors and Editors
Edited by Janet S. Dodd
264 pp; clothbound ISBN 0–8412–0917–0; paperback ISBN 0–8412–0943–X

The Basics of Technical Communicating
By B. Edward Cain
ACS Professional Reference Book; 198 pp;
clothbound ISBN 0–8412–1451–4; paperback ISBN 0–8412–1452–2

Chemical Activities (student and teacher editions)
By Christie L. Borgford and Lee R. Summerlin
330 pp; spiralbound ISBN 0–8412–1417–4; teacher ed. ISBN 0–8412–1416–6

Chemical Demonstrations: A Sourcebook for Teachers,
Volumes 1 and 2, Second Edition
Volume 1 by Lee R. Summerlin and James L. Ealy, Jr.;
Vol. 1, 198 pp; spiralbound ISBN 0–8412–1481–6;
Volume 2 by Lee R. Summerlin, Christie L. Borgford, and Julie B. Ealy
Vol. 2, 234 pp; spiralbound ISBN 0–8412–1535–9

Chemistry and Crime: From Sherlock Holmes to Today's Courtroom
Edited by Samuel M. Gerber
135 pp; clothbound ISBN 0–8412–0784–4; paperback ISBN 0–8412–0785–2

Writing the Laboratory Notebook
By Howard M. Kanare
145 pp; clothbound ISBN 0–8412–0906–5; paperback ISBN 0–8412–0933–2

Developing a Chemical Hygiene Plan
By Jay A. Young, Warren K. Kingsley, and George H. Wahl, Jr.
paperback ISBN 0–8412–1876–5

Introduction to Microwave Sample Preparation: Theory and Practice
Edited by H. M. Kingston and Lois B. Jassie
263 pp; clothbound ISBN 0–8412–1450–6

Principles of Environmental Sampling
Edited by Lawrence H. Keith
ACS Professional Reference Book; 458 pp;
clothbound ISBN 0–8412–1173–6; paperback ISBN 0–8412–1437–9

Biotechnology and Materials Science: Chemistry for the Future
Edited by Mary L. Good (Jacqueline K. Barton, Associate Editor)
135 pp; clothbound ISBN 0–8412–1472–7; paperback ISBN 0–8412–1473–5

For further information and a free catalog of ACS books, contact:
American Chemical Society
Distribution Office, Department 225
1155 16th Street, NW, Washington, DC 20036
Telephone 800–227–5558